Technische Universität München
Lehrstuhl für Datenverarbeitung

Trust between Cooperating Technical Systems

With an Application on Cognitive Vehicles

Walter Bamberger

Vollständiger Abdruck der von der Fakultät für Elektrotechnik und Informationstechnik der Technischen Universität München zur Erlangung des akademischen Grades eines

Doktor-Ingenieurs (Dr.-Ing.)

genehmigten Dissertation.

Vorsitzender: Univ.-Prof. Dr. sc. techn. Gerhard Kramer

Prüfer der Dissertation:

1. Univ.-Prof. Dr.-Ing. Klaus Diepold
2. Prof. Dr. Sandra Zilles, University of Regina, Kanada

Die Dissertation wurde am 29. Januar 2013 bei der Technischen Universität München eingereicht und durch die Fakultät für Elektrotechnik und Informationstechnik am 24. Juli 2013 angenommen.

Walter Bamberger. *Trust between Cooperating Technical Systems. With an Application on Cognitive Vehicles.* Tredition, Hamburg, 2014.
You can obtain various flavours of this book: paperback (ISBN 978-3-7323-0946-7), hardcover (ISBN 978-3-7323-0947-4), e-book (ISBN 978-3-7323-0948-1), free PDF document for open access (http://nbn-resolving.org/urn:nbn:de:bvb:91-diss-20130724-1129245-0-2).

Published by tredition, Hamburg, Germany, http://www.tredition.de.
Cover design by Christian Ullermann, http://www.der-kleine-buecherladen.de.

This book features research within the Fidens project at the Institute for Data Processing, Technische Universität München (TUM), Munich, Germany. For accompanying information and new research visit http://www.ldv.ei.tum.de/fidens.

Key words: cognitive system, cooperation, Dirichlet process, infinite hidden Markov model, interpersonal trust, relational dynamic Bayesian network, reliability, reputation, requirements, self-organising system, social science, society, time-varying behaviour, transfer learning, trust model, vehicle, vehicular ad hoc network, VANET.

Abstract

Researchers from the social sciences and economics consider trust a requirement for successful cooperation between people. It helps to judge the risk in situations, in which a person has the choice to rely on another one. In the future, technical systems will face similar situations. Assume for example that, at a large logistics centre, a robot should reload goods of a ship in cooperation. In the beginning, it must find the right partner out of a set of diverse other robots. To do this selection efficiently without exaggerated security mechanisms, the robot needs trust. Here I consider trust a mechanism, which estimates the certainty of the outcome of the partner's actions.

This dissertation formalises trust between technical systems to set the theoretical foundation for the above idea. It reviews the socio-scientific and technical literature and identifies generic requirements for the mechanism trust. Based on the requirements and further considerations, it presents a conceptual, implementation-independent framework. This new framework, called the Enfident Model, incorporates various facets of trust in form of sub-models. Amongst others, it regards the temporal development of cooperation, the dependency on the task and bargaining, time-varying behaviour of the cooperation partner, learning from experiences, logical constraints of the present situation, and transfer learning to handle unknown situations. With these manifold features described on a conceptual level, the Enfident Model captures existing trust procedures and is suitable for designing new ones. The theoretical part is complemented with algorithms for prototyping trust in individual applications. These algorithms use statistical relational learning to combine logic, learning, clustering and statistics for trust development. They work on a relational dynamic Bayesian network.

Since trust is a social phenomenon, the evaluation features a virtual society of vehicles. These systems cooperate by exchanging information in a vehicular network. They use a trust algorithm to distinguish correct from incorrect information. The simulation shows that the identified trust requirements and the Enfident Model lead to intuitive and consistent results.

Contents

10 Conclusion   **213**

1 Introduction

In a future with many self-organising systems, socio-scientific issues also apply to the society of those machines. Imagine, for example, a future scenario of robots at a large construction site. They have different shapes and abilities as they have been optimised for different purposes, like moving big and heavy items, or cutting and screwing. Some of them have worked together before; others do not know each other, because they are new or belong to different companies.

In this scenario, various complex tasks can only be executed jointly by a group of robots. Imagine a robot has got the job to carry out such a task. It looks for partners, asks them whether they would be willing to do the job, and finally performs the task with their support. Selecting the right cooperation partners is important for an optimal outcome: The partner could have insufficient abilities, be partly defect, or be manipulated to sabotage the task. Thus the organising robot should select those partners, with which it expects to gain the best outcome. This is where trust comes in.

1.1 Problem Statement

The scenario above is an example for the problem this dissertation addresses. The general setting consists of a system that wants to cooperate with another system or a group of systems. Here I understand cooperation as any form of relying on the action of another party. That setting is related to various subjects like reputation, identification of the partner, individual trust development, decision making, reciprocity and information security (see Figure 1.1 on page 3). This dissertation picks out just one. It focuses on the single problem: *How can a system that wants to cooperate with another sys-*

tem or a group of systems predict the cooperation outcome? This prediction should have the form of beliefs in or likelihoods for all possible cooperation outcomes.

The problem can also be considered from another point of view: If a system can predict cooperation outcomes, it has a certain model of the other's manifest behaviour. It cannot look into the other system to see how that system really works. But it can obtain a limited idea of how the other system works just from observing its behaviour over several interactions. This idea is a model of the other's manifest behaviour regarding cooperation. So the problem treated in this dissertation can also be formulated as: *How can a system learn a model of other systems' cooperation-related behaviour?*

I call a mechanism, which can learn this, *trust between technical systems.* The term trust has different meanings in different fields. To address this fact, the next chapter introduces the views of some researchers in social sciences, cryptology and the field of multi-agent systems. It relates them to the problem described above to clarify why I use the term trust in this dissertation. Finally it defines some trust-related terms for the present work. Chapter 4 summarises the state of the art for technical trust mechanisms. The contribution of this dissertation beyond the state of the art is compiled in Section 1.3.

More specifically, this dissertation does not try to simply solve the described problem with a certain algorithm for a specific application. Instead it collects requirements for a trust mechanism in general and derives a conceptual trust model from them. To realise and evaluate this model, an exemplary algorithm is presented. More implementations of the model and optimisations are subject to future research.

In the remainder of this section, I further detail the problem and delimit it from selected other problems that the reader may possibly think of. For this, Figure 1.1 gives some orientation. The term cooperation is interpreted very widely in this document. It includes delegation and all sorts of relying on another party. Consider, for example, a driver that is overtaking another car on a highway. The situation seems free of risk as no third car is around. But still, each of the drivers relies on the other one not to hit the own car (for whatever strange reasons). This situation features a form of loose re-

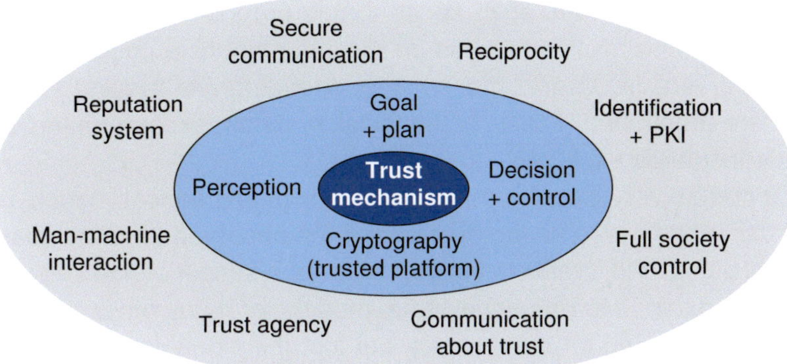

Figure 1.1: This figure shows some mechanisms the reader may think of when talking about trust. The blue ellipse contains modules that work on the individual level. Those in the grey part are used for the interaction with systems: the society level. This dissertation only treats the trust mechanism marked in dark blue.

liance without any explicit agreement. In this dissertation, I still consider it an implicit form of cooperation, as it constitutes a trust situation.

Furthermore the systems here should cooperate without human support. Especially they should develop trust on their own. This is in contrast to systems that use humans as trust sources like classical online reputation systems. That points to an important pre-requisite: In this thesis, a trusting system must be able to assess all facets of a cooperation outcome. Only then, it can learn the cooperation-related behaviour of others on its own.

Related trust methods often include mechanisms for decision making, reputation building, reciprocity enforcement as well as cryptographic data and platform security. I focus on the trust development in the individual and omit society-level features like cryptographic network protocols or reputation building. Moreover I consider decision making and also reciprocity to be different from trust development (see Chapters 2 and 3).

So I propose a mechanism, which just learns a model of the other's behaviour. All the tools mentioned above are related to trust and important for a trusting society. Figure 1.1 depicts this. But they are different from a trust mechanism in the strict sense that is proposed in this dissertation. Moreover the dissertation focuses on machine-machine interaction without human intervention. Every time, when cooperating and trusting systems or agents are mentioned, I refer to technical systems, except if interpersonal trust is considered explicitly.

What comes very close to a trust mechanism is a sensor model. Such a model describes how a sensor transforms the observed physical quantity in an output signal. So it reflects the behaviour of a sensor. A trust mechanism goes beyond this. It learns behavioural models for many other systems, not just one sensor, and for many tasks, not just the single task of obtaining a certain physical quantity. In addition, these other systems are unknown in advance and their basic way of functioning may vary. Still the trust mechanism should provide accurate expectations, even if only few experiences have been made with the other systems before. Thus the trust mechanism must be able to learn various behavioural models; it must be generic. And it should involve transfer learning to quickly adapt to new situations.

The next section introduces various scenarios in which a technical form of trust is useful. The scenarios show that the present work has relevance for the research on cognitive systems, multi-agent systems, sensor networks, vehicular networks and – to some extent – on cryptology; it features techniques from the field of statistical relational learning.

1.2 Motivation and Applications

Trust is only a minor subject in the development of today's technical systems. In contrast to this, interpersonal trust is considered important for personal relationships as well as business organisations (see Section 3.3 and, e.g., Gennerich, 2000, pp. 10–12 for an overview). It improves communication and cooperation, and it is considered a pre-requisite of efficient work flows in groups. If it is so important for people, why is it used only rarely in technical systems? The main reason might be that trust is especially necessary for cooperation between self-organising agents. Strictly controlled work flows, as they are typical today for machine-to-machine interaction, make trust needless. But the proposed idea is important for systems that cooperate in a self-organised way. Such systems will need a trust mechanism to handle the uncertainty when relying on other systems. As a consequence, the reader should venture a glimpse into the future to find application scenarios for trust between cooperating systems.

I use the following exemplary scenarios throughout this dissertation. The first is the scenario of a construction site as described in the previous section. It is similar to the scenario of a large logistics centre with various kinds of robots that cooperate to reload goods from a ship. In both scenarios, the cooperation helps to extend the physical capabilities or to perform tasks more efficiently. In the third example, future cognitive vehicles are driving around while perceiving their environment. To extend their perception range, they exchange all sorts of information, which some vehicles have perceived before. With this form of cooperation, they can efficiently maintain a model of their surrounding world (like a map or a model of the traffic situation) and advise the driver (e.g., where to go or what to give attention to). The fourth scenario features virtual agents at a virtual market place, which trade with each other. So they cooperate as substitutes of persons. These scenarios should give the reader the feeling that trust is helpful for future self-organising systems.

In general, trust supports the following reasoning tasks that appear when cooperating:

1. Select a cooperation partner from several possible ones;

2. Decide whether to cooperate or not if there is a choice not to cooperate at all;

3. Know about the weaknesses of a certain act of cooperation and take their consequences into account;

4. Decide about the correctness of received information;

5. Decide whether the received information about a certain subject is sufficient; and if not,

6. Decide whom to ask for a further opinion about the subject (which is related to Item 1); and finally,

7. Decide whether to accept a cooperation request of another party (which is related to Item 2). So trust is usually needed by both, the one that asks for cooperation and the other one that is asked.

In summary, a self-organising cooperating system needs trust to decide on "how, when, and who to interact with" (Ramchurn et al., 2004, p. 3).

The reader can find many scenarios, in which future technical systems could perform the above reasoning tasks. To support this, I give an overview of the various forms of cooperation, which can be expected in the future (based on Hirche, 2010). It was proposed in CoTeSys, a cluster of excellence of the Deutsche Forschungsgemeinschaft (German Research Foundation), which investigates cognitive systems.

- Two systems interact solely through the environment during the cooperation.

- Two systems share single components and couple one another via information exchange
    - to extend the perception range (joint perception),
    - to extend the physical capabilities (joint manipulation),
    - to increase the learning performance (joint learning), and
    - to find good and efficient strategies for the task execution (joint planning and decision making).

Figure 1.2 illustrates how two cognitive systems can share various components directly.

Both main forms of cooperation can also be mixed. Trust is helpful in all cases.

With this schema of cooperation forms, the reader might get an idea of the various applications we can expect of future self-organising systems. The previous list of reasoning tasks shows that a trust mechanism can strongly

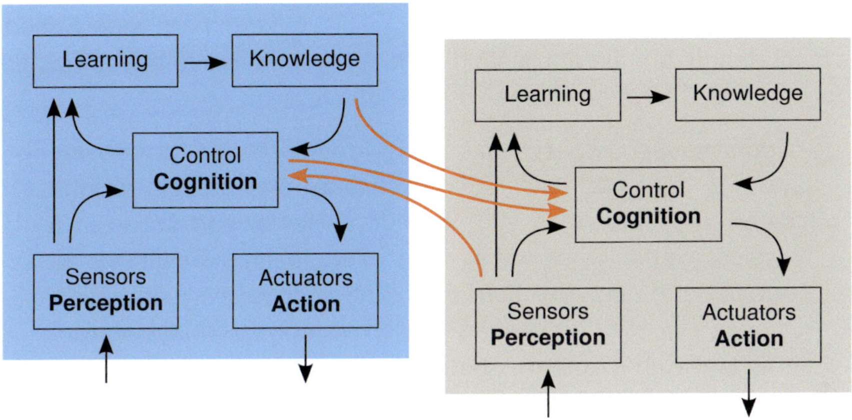

Figure 1.2: Examples of how two cognitive systems can share their components (based on Hirche, 2010). The black lines indicate data flows between the components in one system. The orange lines refer to data flows, which are realised by communication between two different systems.

support the reasoning in these applications. So there is a wide range of use cases trust can be applied to. But is trust really necessary or could it be substituted with better planning and control in the scenarios? Full control over complex situations with several interested parties is difficult and, thus, expensive. Imagine, for example, a large harbour in the future. The robots there belong to different parties, have various ages and come from several manufacturers. So full control is difficult here. Avoiding strict global control is the exact idea behind self-organising systems. Thus trust enables those systems to cooperate efficiently without expensive procedures for security enforcement. This concern is similar to that of Gerck (2002), who recommends trust for the Internet because of its self-organising nature. For him, using trust instead of full surveillance has the advantages of a simpler and more modular system design as well as lower costs.

Above I used the notion of a cognitive system. This kind of system has the ability to trust, because it can perceive and understand its environment in order to judge past acts of cooperation and to learn from them. And this kind of system has a need for trust, because it should engage in cooperation and reason about cooperation. Therefore cognitive systems are widely used in this dissertation, but the application of trust is not limited to them. This term is defined in CoTeSys as follows:

> *"Cognitive technical systems (CTS)* are information processing systems equipped with artificial sensors and actuators, integrated and embedded into physical systems, and acting in a physical world. They differ from other technical systems as they perform *cognitive control* and have *cognitive capabilities.* *Cognitive control* orchestrates reflexive and habitual behavior in accord with longterm intentions. *Cognitive capabilities* such as perception, reasoning, learning, and planning turn technical systems into systems that *'know what they are doing'."* (Buss et al., 2007, p. 25)

1.3 Contribution of This Dissertation

This dissertation has the objective to improve the understanding and modelling of trust between cooperating technical systems. To achieve this, it contributes the following to a theory of technical trust.

It discusses the term and mechanism "trust" across disciplines and introduces research on interpersonal and technical trust to compare various views. In contrast to the state of the art (e.g. Castelfranchi and Falcone, 2010; Engler, 2007; Kassebaum, 2004), this dissertation presents *interpersonal trust as an input-output system.* This new view makes it easier to relate trust between persons and between machines with each other. In addition, the presented interdisciplinary discussion is deeper than the state of the art. This leads to a different understanding of technical trust, especially regarding the following questions: What notions of trust can be distinguished (Section 2.5)? How does trust differ from related mechanisms (Chapters 1 and 2)? What influences trust development (Sections 3.2 and 6.2)? How do interpersonal trust and inter-machine trust differ from one another (Sections 3.4 and 10.2.5)? This work results in clear, well-founded technical concepts for different notions of inter-machine trust. It is necessary, because the present state of the art lacks a sufficient theoretical framework for the trust model presented in this document.

The interdisciplinary research together with an analysis of future trust scenarios leads to a formalisation of trust between technical systems. This formalisation is the core contribution of this dissertation. It consists of *general application-independent requirements* on a trust algorithm and a *conceptual implementation-independent model of trust*. The requirements are postulated together with a review of the technical literature in Chapter 4. Formal requirements for a trust mechanism are unique in the literature. While some authors (e.g. Ramchurn et al., 2004) review the literature on trust, they do not derive requirements from it. Furthermore the new conceptual model of trust describes various aspects of trust development and can be understood as a meta-model to create new application-specific trust algorithms. It is presented in Chapter 6 and called the *Enfident Model*. The following list details its main features with a focus on those that are rarely found in other trust models.

- The Enfident Model *evaluates a trust situation comprehensively.* It explicitly names three aspects: the cooperation partner(s), the cooperation agreement and the task to fulfil. It combines them as entity classes in a relational sub-model; each of the entity classes groups several attributes of the trust situation. Present trust models consider the attributes of one or two of those entity classes only, as Section 4.2 points out.

- This relational sub-model can reunite two lines of research on technical trust, which are detailed in Section 4.2. Today, most trust algorithms rate previous cooperation outcomes and derive trust from these ratings. In contrast, the socio-cognitive trust models derive trust from beliefs about the cooperation partner in the given trust situation, basing their theory on belief-desire-intention agents. *These beliefs can be located in the Enfident Model in the same way as the cooperation outcomes and contextual information.*

- Section 4.4.1 shows that some trust algorithms base their outcome prediction on *past experiences*, while others use *logical constraints of the present situation.* The Enfident Model addresses both information sources. This is unique in the literature.

- Most trust algorithms just rate the act of cooperation. In contrast, this dissertation *makes the cooperation outcome the first class object.* The subjective likelihoods of the possible cooperation outcomes (named the trust distribution) should be predicted directly and as complete as possible. If necessary, a rating can be derived from them in a subsequent step, either in the trust algorithm or in a decision algorithm. The trust algorithm in ElSalamouny et al., 2010 is one of few examples that put out the cooperation outcome instead of a rating.

- Present trust algorithms compute specific trust for a certain purpose. The needs of a reputation system, for example, or the trust problem of an autonomous agent define that situation. The literature of the social sciences shows though that people can express trust for all sorts of attribute combinations like: the trust in a certain cooperation partner or the trust regarding a certain situational setting (e.g. meeting at night) (see Section 3.4.1). The Enfident Model resembles this with the *concept of querying.* This concept is unique in the technical literature. It enables a system to compute trust for a specific trust situation or to exchange the trust in various objects with other systems – with just one single trust model.

- The Enfident Model explicitly *models trust-related changes in the mentioned entities over time.* For example a cooperation partner could change its behaviour, which means its internal way of working, because of defects or software updates. I found a related functionality only recently in the literature: ElSalamouny et al. (2010) model the time-varying behaviour of a single cooperation partner as a hidden Markov model. The Enfident Model includes similar sub-models for all entity types not just the cooperation partner and entangles those sub-models across entities. Moreover the Enfident Model proposes a time-dependent likelihood for the state transitions.

- Trust develops over an ordered sequence of acts of cooperation. An act of cooperation may in turn consist of an ordered sequence of interactions. The trustor can evaluate trust at any time during an act of cooperation. Some information may be known at that time, other information may be unknown and some information may change from interaction to interaction. To my knowledge, no present work contains such a *comprehensive sub-model for the temporal development during a single act of cooperation.*

- A trust mechanism should help to handle new, uncertain situational settings. Therefore it must *transfer knowledge* from other, even different settings to this new one by utilising similarities (Pan and Yang, 2010). Rettinger et al., 2008 is the only present work that realises this functionality satisfyingly.

The Enfident Model combines all these features in a coherent model and shows how they can interplay with each other. Present trust models focus on few of them only. This listing also clarifies why the Enfident Model can serve as a meta-model to analyse existing trust algorithms.

To realise this functionality, I propose *algorithms that combine clustering, learning, logic and probability theory in a relational dynamic Bayesian network* (e.g. Manfredotti, 2009). They are based on the algorithms in Xu, 2007 for static relational Bayesian networks and the algorithms in Van Gael, 2011 for infinite hidden Markov models.

For the evaluation, the Enfident Model is applied to the scenario of cooperating cognitive vehicles. This scenario features a whole "society" of self-organising systems. Since trust addresses a social problem, the evaluation with a realistic technical society matches best here. To my knowledge, such an evaluation is unique in the literature and was a complex undertaking.

1.4 Organisation

The organisation of this dissertation uses a methodology that follows the phases of a systematic engineering process with use cases, requirements, design, implementation and testing. At the same time, the text is organised in two parts: a generic and an application-specific part. To avoid duplication of text, some phases of the above process are detailed in one part or the other only, as described in the following.

Problem definition and use cases. Chapter 1 introduces the problem and sketches application scenarios. Chapter 2 then compiles views on trust from various fields to find a definition of trust and related terms for this dissertation. Those views and the definitions further clarify the problem. A comprehensive description of a single application together with use cases can be found in Chapter 8.

Requirements. Chapter 4 presents the requirements. They are based on a review of the socio-scientific literature on interpersonal trust in Chapter 3 and of the technical literature on trust in Chapter 4. Own considerations complement them.

Design. The requirements lead to an application- and implementation-independent design of a trust mechanism: the Enfident Model (Chapter 6). Chapters 4 and 6 together show that the Enfident Model suits as a framework to analyse existing technical trust algorithms and to design new ones. The preceding Chapter 5 introduces the notation of some mathematical tools that are used throughout the remainder of this document.

Implementation. Chapter 7 proposes implementation techniques for the Enfident Model. These techniques originate from statistical relational learning and are just implementation examples, because other techniques seem reasonable as well. Chapter 7 marks a first step towards a concrete algorithm. However the attributes are still unknown; they depend on the application. Chapter 8 then applies the model to a specific scenario. In this step, attributes can be identified and the algorithms can be completed.

Test. Chapter 8 describes the evaluation method. It introduces the application scenario of cognitive vehicles that cooperate through a vehicular network and defines the simulation environment. The evaluation results and the discussion are combined in Chapter 9, but separated in the subsections. In this way, one subject can be evaluated and discussed in one place, while the reader can still distinguish the results and their discussion.

Chapter 10 summarises the dissertation. For this purpose, it also relates the Enfident Model back to selected findings from social sciences. Finally it points out directions for future research.

2 Clarifying the Concept of Trust

Trust is a term of everyday speech. People know it and have formed it during the integration in her linguistic environment. As a consequence, the meaning of the term varies between individuals – but also between researchers on trust. Various disciplines investigate trust and even within a field, people have a different understanding of what trust is. In contrast, a central term of a scientific paper should have a clearly delimited meaning.

As a consequence, I introduce conceptualisations of trust from different disciplines in this chapter. Because trust is primarily associated with humans, the view of social scientists is discussed first. Because interpersonal trust serves as a prototype for the trust concept in other disciplines, it is discussed more comprehensively than the other trust concepts.

Interpersonal trust is a mechanism that has not been invented for a special aim, but simply found to be there. Therefore some scientists have argued on its purposes. Their considerations are introduced in Section 2.2. Some of the purposes the same problem as that mentioned in the introduction. This is the reason, why I speak of trust between technical systems: This technical trust mechanism should provide a similar functionality as interpersonal trust, although both mechanisms might work differently.

Sections 2.3 and 2.4 cover the concept of trust in the technical fields of cryptology and multi-agent systems. Finally, Section 2.5 introduces a definition of trust between technical systems in the form it underlies the remaining dissertation.

2.1 Interpersonal Trust

In the literature, many authors choose their own definition for interpersonal trust. Often these definitions are operationalisations with only a limited applicability (Narowski, 1974). In order to represent a construct that is subject to investigations, the concept must describe something observable. These observable criteria constitute an operationalisation of the term then.

This section describes the concept of interpersonal trust (German: *zwischenmenschliches Vertrauen* or *interpersonales Vertrauen*) as an attempt to integrate considerations from different authors. To avoid just another new definition of the concept, that of Kassebaum (2004) is taken. It integrates many definitions of the literature. Especially it incorporates the affective, behavioural and cognitive component of trust; many other authors considered only some of them (Narowski, 1974, p. 125). However it is hardly possible to come to a common understanding of trust between you as the reader and me as the author within three sentences. For this reason, I highlight key aspects of the definition afterwards.

> "Interpersonal trust is an expectation about a future behaviour of another person and an accompanying feeling of calmness, confidence, and security depending on the degree of trust and the extent of the associated risk. That other person shall behave as agreed, not agreed but loyal, or at least according to subjective expectations, although she/he has the freedom and choice to act differently, because it is impossible or voluntarily unwanted to control her/him. That other person may also be perceived as a representative of a certain group." (Freely translated from Kassebaum, 2004, p. 21)

Most parts of the definition describe the so called *trust situation*, in which someone reasons about the behaviour of another one. Both persons may tightly work together or be loosely coupled according to the definition. This is the wide understanding of cooperation that is underlying this dissertation as already mentioned in the introduction. The term cooperation is still chosen because it emphasises the relational aspect of trust between two systems.

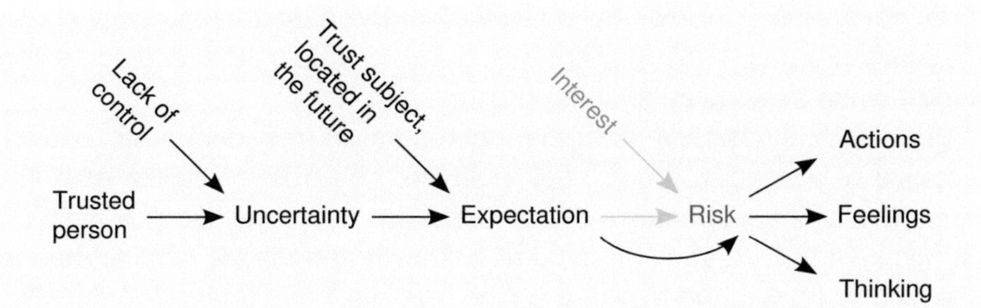

Figure 2.1: Key aspects of the interpersonal trust definition. A trust situation involves an object to trust, the trusted person, and uncertainty about a trust subject in the future. The trusting person forms an expectation about the outcome of the trust situation. For some authors, the possible outcomes need to involve risk, for others they just may do so. The formed trust attitude can in the end result in actions (the behavioural component), feelings (the affective component) and thinking (the cognitive component).

But Kassebaum goes beyond the definition of a trust situation. He also emphasises that interpersonal trust is an expectation and a feeling. As an attitude, trust expresses in affection, behaviour and cognition. The affective component can be considered one difference between interpersonal trust and trust between technical systems.

In the following, key aspects of the definition are detailed and discussed with regard to the literature. Figure 2.1 visualises them.

The trusted person. Interpersonal trust involves two parties who interact with each other: On the one hand side, there is the person who trusts, ego, the trustor (German: Vertrauender or Treugeber). In this document, I use the name Paula (P) for this person in many examples. On the other hand side, there is the person whom is trusted, alter, the trustee (German: Vertrauensperson or Treuhänder). I name this person Oliver (O). In the case that mutual trust develops over time in many interactions, both parties are ego and alter at the same time.

As part of the interaction, ego judges alter to be trustworthy or untrustworthy. Such a judgement about alter's traits and motives is called attribution in social psychology. Studies have shown that the attribution process is very subjective (Forgas, 1985, p. 77). This is a basic finding that should be kept in mind when thinking about interpersonal trust.

The actor can perceive the other person as an individual or as a representative of a specific group. For example, one trusts a police man in a dangerous situation, because this person holds the role of a police man, but not because this person is trusted as a known and maybe familiar person. This kind of trust in the role of the other is called *role trust* (German: *Rollenvertrauen*) by Strasser and Voswinkel (1997). In showing trust in the role, the trust in the abstract system of the police becomes practical. Therefore a person can have trust in the working of a system. Luhmann (1979, Chap. 7) calls this type of trust *system trust* (German: *Systemvertrauen*). It is important for a complex society with a high degree in the division of labour. Trust can be established between two persons, unfamiliar with each other, but acting on behalf of a trusted system. Gennerich (2000, pp. 40–44) extends this concept to general social groups to which a person can manifest a social identity. For example, fans of a soccer team form a community within that they trust each other to a certain extent. In contrast to system trust, Luhmann (Chap. 6) calls the trust in an individual – which is mostly based on familiarity – *personal trust* (German: *persönliches Vertrauen*).

Note that the object of trust can also be a thing or my self (self-confidence). These forms of trust are out of the scope of this thesis, as they are not referred to as interpersonal trust.

Lack of control, complexity and uncertainty. The trusted person must be free to some extent to behave trustworthy or untrustworthy (Kee and Knox, 1970). This freedom may be forced by the situation or voluntarily given by the trustor. From the point of view of the trusting person, it is a lack of control over the situation that forces to trust.

For Luhmann, this is an important feature that characterises that kind of complexity the trust mechanism addresses: It is "that complexity which enters the world in consequence of the freedom of other human beings" (Luhmann, 1979, p. 30).

The lack of control can also be understood as a lack of knowledge. Trust is a "middle state between knowing and not-knowing" about another person (Simmel, 1968, p. 263). Luhmann (1979, Chap. 2) details this in the following way. If Paula knows how Oliver will act and how the cooperation will end up, she can make a rational decision and needs not trust. If she knows nothing about the specific problem, she cannot trust but only hope. The trust decision forces the trustor to choose one out of the many possible scenarios the future offers. Altogether the lack of control and knowledge results in an increased uncertainty about the future and, thus, in an increased complexity that is inherit to a trust situation. It comes from the trusted person being there and free to act. Trust is a mechanism to cope with this uncertainty.

Some authors think that this mechanism is an irrational process only partly based on clear evidence. It incorporates some rational decision calculus about the uncertainty but deliberately goes beyond that. "Trust always extrapolates from the available evidence" (Luhmann, 1979, p. 26). Kee and Knox speak of a "subjective probability" and an inner, not rational "certainty or uncertainty about O's trustworthiness" (1970, p. 359). This irrational process is driven by wishful thinking (Koller, 1997; Oswald, 1997). Section 3.4.3 summarises some of the irrational findings about interpersonal trust.

In contrast to this, users expect a technical system to act predictable and rational. So while both, persons and technical systems, need trust to handle the uncertainty, the way they do it might be different.

Subject of the trust situation and expectation. Despite the uncertainty, Paula must still act, either by relying or by not relying on Oliver. For this, she forms an expectation about the future. Burt and Knez propose "Trust is anticipated cooperation" (1995, p. 257) as a compact definition of interpersonal trust. Luhmann emphasises the anticipation of the future, as well: "To show trust is to anticipate the future. It is to behave as though the future were certain" (1979, p. 10). The future consists of many possible scenarios; only one can become present – a process of complexity reduction. Someone who trusts chooses from all the possibilities of future presents. With this choice, the trusting person simplifies her internal future.

At the same time, Paula looks back into the past too. She uses prior experiences to form an expectation about the future. This is detailed below.

The expectation also specifies what to expect, the cooperation subject of the trust situation. "Trust therefore always bears upon a critical alternative" (Luhmann, 1979, p. 24). Note that this statement already points to the proposition of Requirement 2 that a trust mechanism should put out a probability distribution or a belief mass distribution over all possible future worlds (see page 79 of this document).

Paula is only able to build up a clear expectation for one of the future worlds, if Oliver acts predictable. So the attribution of predictability and consistency to Oliver is a key requirement to establish trust (Gennerich, 2000; Rempel et al., 1985). Paula takes her collected experiences from the past and transforms it to an expectation for the future. Trust is thus based on social learning (Blomqvist, 1997, pp. 280 and 283).

Risk. Luhmann restricts that not every expectation is trust-related. Expectations of trust are "only those in the light of which one commits one's own actions and which if unrealized will lead one to regret one's behaviour" (Luhmann, 1979, p. 25). Thus the individual must have some interest in the outcome of the trust situation. This interest corresponds to a value. And because of the uncertainty the value is at risk. In addition to the expectation above, risk incorporates a value because of own interest.

Many authors support this restriction. Some others negate it though. For example, Jones (2002), a researcher from the field of informatics, criticises: "While it is true to say that a goal-component of this sort is often present, this is by no means always so. For example, x might trust y to pay his (y's) taxes [...], even though it is not a goal of x that y pays" (p. 229). He regards trust as an expectation towards another one without the need of own interest. Thus a trust situation needs or needs not involve risk. Figure 2.1 depicts this with the additional arrow that bypasses the term risk. Note that the concept of interpersonal trust can merely be observed only. In contrast to this, trust between technical systems is designed. So whether to include the risked value in the trust computation, is a design decision. It is discussed in Section 2.5.

Of what kind is the risked resource? It may be a material resource resulting in a direct financial harm, but also time, effort, and trouble. Rempel et al. give a couple of examples mostly relevant in intimate relationships: "[...] trust involves a willingness to put oneself at risk, be it through intimate disclosure, reliance on another's promises, sacrificing present rewards for future gains, and so on" (1985, p. 96). Gennerich (2000) regards the own identity as a resource that is always at risk, sometimes more, sometimes less. I detail this – in my opinion interesting – thought in Section 2.2.

The harm arises if the other one acts untrustworthy. If he fulfils the trust though, the trustor has a benefit from that. Examples for the benefit are future reciprocity in the relationship, health when going to the doctor, or a monetary benefit when accepting a "good deal". The benefit is also associated with a subjective probability. Both together form a positive risk or chance. Some authors like Luhmann (1979, Chap. 4) require that the perceived risk must be larger than the perceived chance in a trust situation. Otherwise the decision is more rational.

The risk is not only in the situation but also in the motives of the other. The trustor must decide whether the motives he perceives from the other are sufficient to trust in this situation. While Kee and Knox (1970) require that both sides must be cognisant of the risk for the trustor, Strasser and Voswinkel (1997) do not even postulate that the other must behave loyal. In their understanding, it is sufficient that an enemy or competitor behaves as expected. Then the risk for the trustor may be hidden towards the other.

The investigation of Koller (1988) supports the assumption that the degree of trust depends on the risk.

Heisig (1997, pp. 131–133) emphasises the binding effect of trust. The trusted person usually perceives a high obligation to fulfil the request. The act of trusting is an appreciation towards the trusted person, a moral donation. Similarly complying with the act of trust also has a high moral value. Betraying the trust would reject the appreciating donation. Even without formal sanctions, a betraying person would be stigmatised in his social environment. In fact, not only the trusting person risks her own identity, but also the trusted person. For this reason, Heisig thinks that the risk is too strongly emphasised in the literature about trust.

The affective, behavioural, and cognitive component.   Trust is an attitude towards the trust situation, including the trustee. Human beings can express an attitude, and hence trust, in three ways: with an affective, a behavioural, and a cognitive response (Fazio and Petty, 2008, pp. 7–11).

The affective response of trust consists of "feelings of confidence and security" (Rempel et al., 1985). The behavioural component can be formulated in a compact way as: "Trust is anticipated cooperation" (Burt and Knez, 1995, p. 257). Within this component, trust is expressed as a verbal intention or as an action with cooperative tendency. As the cognitive response, the individual reasons about the uncertainty in the situation and the attributes of the other (see Chapter 3.2).

Narowski (1974, pp. 125–130) points out, that the definitions in the literature diverge, mainly because they contain only few of these three responses. But trust is reflected in all of them together.

Summary.   The concept of interpersonal trust features two persons who are connected through the expectation of the one person about the other person's behaviour. So trust is an attitude towards the behaviour of another person in a given situation. This situation is referred to as the trust situation. As a consequence, the trust attitude depends on three independent variables: The one who is trusting, the one whom is trusted and the situational context. This is a main result with consequences to the modelling of interpersonal trust in Chapter 3 as well as trust between technical systems in Chapters 4 and 6.

The attitude represents an expectation about the future state of the world. So it assigns beliefs or subjective likelihoods to some of the possible states. It is debatable whether trusting includes the interest of the one party in the other's behaviour. If so, the trust situation would be characterised by perceived reliance and risk.

Some authors argue that the trusted person must be aware of the risk the trusting person is exposed to. This leads to an increased obligation of the trusted person not to betray. Other authors consider the awareness as a possible advisory phenomenon but not as a pre-requisite for trust.

Finally trust is obtained in response to the current situation with a process that is partly rational and partly driven by own desire. The obtained trust becomes manifest in affections, actions (behavioural trust) and thinking.

Section 2.5 summarises how interpersonal trust related to trust between technical systems as proposed in this dissertation – discussing the similarities and differences.

2.2 Functions of Interpersonal Trust

When there is a new effect in the natural sciences or a new technical achievement, it quickly raises the question: What is it good for? This question is interesting regarding the trust construct too. In the literature, I found that the trust mechanism supports three functions; they are described below.

Identifying a function of a construct sounds like finding a solution for a problem. Luhmann (1979, Chap. 4) points out though that trust is accompanied by a problem as well. The complexity is still there and, after all rational considerations, risk remains. Trust is more a substitute problem. So the advantage must be in form of the remaining problem. It seems that human beings can bear a trust problem more easily than the original problem.

Reduction of social complexity. When a mother entrusts her child to a babysitter for an evening, many things can happen during this time: Everything could work fine; the baby could sleep well, while the babysitter reads a book or watches TV. But also, the babysitter could watch a cruel television program with the child. Or, the babysitter could not care about the child, although it screams. The babysitter could even murder the child and run off.

Luhmann defines the concept *complexity* as the "number of possibilities which are opened up through system formation. It implies that the conditions (and hence boundaries) of possibility can be specified, that the world becomes constituted after this fashion and also that the world contains more possibilities than can be realized, so that in this sense it has an 'open' structure" (Luhmann, 1979, p. 6). The example above shows that the future offers many possibilities. This openness of the future makes the situation complex. Interpersonal trust is specifically about that kind of complexity, "which enters the world in consequence of the freedom of other human beings" (Luhmann, 1979, p. 30) – it is about social complexity.

In that the mother trusts the babysitter, she leaves many of the possibilities of the future meaningless. This reduces the perceived complexity and gives the mother the necessary confidence and security to spend a pleasant evening away from home. Things become even clearer, when considering what happens if human beings would not have a mechanism like trust. Luhmann thinks: "A complete absence of trust would prevent him [the human being] even from getting up in the morning. He would be prey to a vague sense of dread, to paralysing fears. [...] Anything and everything would be possible" (1979, p. 4). In a positive sense, leaving some bad possibilities of the future meaningless opens the mind for new experiences. *So trust is an important base for taking new chances.*

Besides the uncertainty and complexity, a trust situation may also involve risk. Then another purpose of the trust mechanism is that it helps to handle the risk.

Risk management.　If the trusting person has some interest in the other's action, if some gain or loss is at stake for her, trusting is associated with risk taking. So perceiving the risk helps to act correctly.

Following exchange theory, the expected risk could be quantified as the product of the potential loss and the probability of betrayal; analogously the expected gain would be the product of the potential gain and the probability of fulfilment. A rational agent would then compare both, expected risk and gain, for a decision (Coleman, 2000, Chap. 5).

In Koller's (1997) model, trust depends on the perceived risk and the perceived importance of the interaction goal. The higher the importance, the higher is the maximal accepted risk. This model can explain irrational findings like those in Section 3.4.3. It emphasises the individual perception of the trust situation by the trusting person. Trusting is also wishful thinking. So the trust mechanism manages the risk in balance with the importance.

Trust can handle the risk also in another way. In many situations, the trust situation is also risky for the trusted person. A betrayal of trust would lead to sanctions of the community like stigmatisation. No in such situations where both are cognisant of the risk for the trusting person and the trusted person, the risk is indeed reduced (Heisig, 1997, p. 133). (Compare this also with the note at the end of this section.)

Besides the complexity reduction and risk management, trust has another purpose that is considered only rarely. The risked resource may be the own identity. Then risk management is actually a identity building as the following thoughts show.

Preservation of the own identity. I reconsider the above mentioned example of a mother that entrusts her child to a babysitter. If the babysitter does any harm to her child, she will perceive herself as a bad mother and think that others will perceive her in the same way. So the babysitter impaired her identity. To clarify this, imagine the process of identity building as a mutual alignment of the image that the mother would like to be seen as (the intended identity, I) and the image that she perceives others see her (the perceived identity, me). (See Gennerich, 2000, pp. 50–57 who follows Mead, 1974.)

According to Gennerich, trust can develop if, with regard to the given situation, the intended role-identity and the perceived role-identity match. When this matching successes across many different situations, the trust in the other one is high, because this is only possible, if the other one understands all intended identities and shows a high variability. Then it is likely that the other one will confirm future intended role-identities as well – he will show the desired behaviour. In contrast, if the matching fails, suspicion and distrust increases. From this point of view, trust is a mechanism to preserve one's own identity. Note that this view also includes an expectation about the other's behaviour.

If Paula, who lent a book to Oliver, does not get back her book, she may ask herself whether Oliver considers her a friend, whether Oliver understands her person with regard to her tidiness, or she thinks everyone mucks around with her. In many situations, the identity is impaired only marginally like in this example. But still, even small trust problems affect the own identity in some way. As another example, the relationship to one's superior affects the occupational identity. In some way, trust problems always raise questions like: "Who am I?", "Who would I like to be?", "Am I good enough?", or "Am I likable?" (Gennerich, 2000, p. 49).

Luhmann (1979, Chap. 4) describes a cognate idea from a systems point of view. The human being counteracts the complexity of the outer environment with inner order leading to perceived confidence. Trust is a mechanism to constitute and stabilise this inner order. It does so in a way that the inner order is more important than environmental evidence to perform the complexity reduction. Thus the trust mechanism is sustained by its importance for the inner order and confidence.

Finally note that Oliver is another ego who needs to develop his identity. Paula's trust in Oliver depends on how she perceives him, that is, what identity Oliver shows towards her. His identity makes Oliver different from others. And because of his identity, Paula could choose Oliver to help her in the trust situation. From Oliver's point of view, these considerations of Paula support or violate his identity. And in the distinct way how Oliver fulfils Paula's request, Oliver exhibits his identity. So for Oliver, accepting a trust request is a problem of self-expression and identity building; and therefore, the request itself can be a donation from Paula for him, if it meets his identity. In the end, Oliver must balance the desire to be trustworthy with other aspects of his personality (Luhmann, 1979, Chap. 8).

2.3 Trust in Information Security

Classical books about security in IT systems (e.g. Menezes et al., 2001) use trust very frequently in expressions like "trusted third party", "trust in a sever" or "trust chain". But nowhere, the term is defined. From reading, it turns out that the user or the system can either completely trust or not trust at all; there is no degree of trusting. Often trust is used in the context of public key infrastructures and is established through authentication.

Standards related to this context try to be more general. For example, the ITU standard X.509 defines: "Generally, an entity can be said to 'trust' a second entity when it (the first entity) assumes that the second entity will behave exactly as the first entity expects. This trust may apply only for some specific function. The key role of trust in this framework is to describe the relationship between an authenticating entity and an authority; an entity shall be certain that it can trust the authority to create only valid and reliable certificates" (ITU, 2008, p. 6). While the first sentence is quite general, the standard focuses on a technical infrastructure that helps to ensure and enforce certain kinds of policies. In the end, developing trust reduces to cryptographic authentication.

So what happens here? The policies and the proposed technical infrastructure have an enforcing character that is necessary, when people want cooperate, although they do not trust one another directly. With regard to this, they are similar to some legal rules and their enforcement. Sitkin and Roth (1993) investigated such legalistic mechanisms in the context of organisations. They point out that these policies often fail to establish trust. Instead they substitute trust with a control-based confidence. The control comes from the enforcing mechanisms that are integrated in the policies and the infrastructure.

Gerck (2002) argues in a similar direction for the context of Internet communication. He compares information security and trust with closed loop and advanced open loop control. Closed loop control realises complete surveillance. In contrast, the open-loop controller actuates without prior observation, just based on estimation. Gerck still allows the open-loop controller to observe some system variables every now and then. This way, it can adapt its model of the controlled system. Trust is similar in the way that the trustful action happens without full control. But the trusting person observes its environment and its partners in order to improve her future decision.

Gerck concludes that introducing trust in Internet communication has similar advantages like open loop control: "simpler systems (hence, better fault-tolerance); immediate response (i.e., nothing needs to be measured in order for it to actuate); easier design (e.g., avoiding probable but unknown pitfalls of complex designs); easier interfacing (i.e., suffers less influence from and also exerts less influence on the rest of the system); modular design (i.e., complete and interchangeable); and less cost" (p. 23).

2.4 Trust in Multi-Agent Systems Research

In the field of multi-agent systems, people usually model trust following the work in social sciences. Ramchurn (2004), for example, refers to Dasgupta, when he defines: "Trust is a belief an agent has that the other party will do what it says it will (being honest and reliable) or reciprocate (being reciprocative for the common good of both), given an opportunity to defect to get higher payoffs" (p. 9). Besides the expectation about the other's behaviour, this definition emphasises two more aspects. First, the trusted agent should act reciprocally to increase the general benefit. Making a society of agents work is an important aim in multi-agent systems. The name of the field already points at the society level. For Ramchurn, reciprocal behaviour is important to reach this aim. Consequently, he added it to his definition of trust. Second, the trusted agent should not only have the chance to betray. Rather it must have a higher advantage in this case. This condition should avoid that it is not just the trusted agent's rational choice that lets the reliance work.

Castelfranchi and Falcone (2010) choose a different approach. They state that a subjective probability or belief about the other's behaviour is insufficient for trust. Instead they derive a content model of trust from the work in social sciences, their socio-cognitive model of trust. They start with defining trust as a relation of five parts ($\mathrm{TRUST}(X, Y, C, \tau, g_X)$): "$X$ trusts (_in_) Y in context C _for_ performing action α (executing task τ) and realizing the result p (that includes or corresponds to her [X's] goal $Goal_X(g) = g_X$)" (p. 36). The trusting agent X should evaluate Y's competence, willingness and unharmfulness positively. In addition, X must have an interest in Y's action, a goal it desires; and it must positively expect that this goal can be reached. Moreover it must count on this positive expectation. This means, it must really consider whether to rely on Y. Only then, the attitude of X is trust. Besides these thoughts about the trusted agent, trust also includes considerations about the opportunities and the obstacles of the task in the context.

Castelfranchi and Falcone distinguish the trust attitude, the decision to trust and the act of trusting. While the trust attitude is the mental counting on Y as described above, the decision to trust goes a step further by considering the trust attitude towards others and deriving an intention to act. The decision is finally realised by the act of trusting, which may be temporarily separated from the decision. All in all, the theory of Castelfranchi and Falcone is rather complex. This section only presents a short glance on it without being complete.

Jones (2002) criticises the narrow definition of Castelfranchi and Falcone. For him, trust needs no goal component; it can be just a positive evaluation. Since Castelfranchi and Falcone allow that the trusting system evaluates the trusted system positively and negatively for reaching the goal, he wonders why the trusting system cannot be indifferent towards the trusted behaviour as well. Jones proposes a simpler content model of trust. The trusting system must have a positive rule-belief and a positive conformity-belief to trust. The rule-belief basically expresses that there is some regularity in the trustee's behaviour or norms that restrict the other's actions. So this belief regards the past. The conformity-belief reflects whether the trusting system expects the trusted system to follow this rule in the future again. So it extrapolates from the past to the future.

In summary, the definitions introduced in this section feature different concepts: The likelihood of the trustee's future behaviour, the own goal, reciprocity and the payoffs of the trusted system. The definition of Ramchurn mostly focuses on the purpose of trust: It is a mechanism that forms a belief. The other two definitions also determine what should be including in the process of belief formation (the belief sources).

The work of Castelfranchi and Falcone is quite elaborate. While it addresses social sciences and multi-agent systems, it disregards differences between the trust of a human and of a technical agent though. In my opinion, this has two weaknesses. First, people may act more arbitrarily and irrational compared to what technical systems are expected to behave. In some of their examples that should support their trust theory, I would say, a person may act like this but a machine should not. (Compare this with the paragraph about uncertainty in Section 2.1 and with Section 3.4.3.) And second, Castelfranchi and Falcone try to explain interpersonal trust. This is a mechanism that is already there and can only be observed, not designed. So they describe a specific implementation of trust: interpersonal trust. In contrast, a technical form of trust is a mechanism that is designed for a certain purpose. As a consequence, a definition of trust would be needed that is based on implementation-independent criteria.

2.5　Definition of Trust in This Dissertation

This section defines *trust between technical systems* by abstracting interpersonal trust. For this, it discusses some arguable aspects of the trust definitions so far and identifies those that are necessary aspects and those that are just accompanying ones. This results in definitions of some notions of trust. They should just describe, what trust is, but not, how it works. I use the term agent instead of technical system in this section to remain in line with other disciplines.

Prediction, interest, decision and action.　When regarding interpersonal trust as an attitude, it refers to an inner state, which expresses as feelings, actions and thinking (the affective, behavioural and cognitive component). For some researchers, it incorporates own interest and, thus, risk. Others just require an expectation about the other's action. The situation is similar in technical fields. All of these notions of trust are different objects that are important for the research on trust but not clearly distinguished in everyday speech. I do not say, one of them is trust and another one is not. For the scientific discourse, they should have different names though. As a consequence, I define names for them below: probabilistic trust, interest-related trust, trust-related decision and trust-related behaviour. No one of these names is just "trust". Instead the term trust may refer to all of them. If the specific meaning of trust is not clear from the context though, I use one of these specific terms.

While the forms of responses to the trust attitude (feelings, actions and thinking) are important to understand and investigate interpersonal trust, they are not necessary for a definition of trust in general. One kind of system like a human being may respond in this way, another kind of system may respond in a different way. Especially I do not require the notion of a feeling for trust. A system may express trust as a feeling but needs not.

Representation of trust. In some contexts, trust can be high or low. Then it refers to a specific action of the other agent. The degree of trust describes, how likely this action is or how relevant to reach a certain goal. For practical reasons, the expectation could refer to several possible actions of the trustee. These can be rated separately but handled as a collection. To distinguish the expectation about a single action from the expectations about several possible actions, I use the term *trust value* for a single rating and *trust distribution* for a collection of ratings. The trust value can be understood as a belief or subjective probability that the specific action will happen. The trust distribution could be realised, for example, as a belief function (Shafer, 1976) or as a (subjective) probability distribution. Section 4.3 discusses with regard to the technical literature on trust, why both representations may be reasonable.

Mechanism and state. Some authors regard trust as an inner state that can be high or low as in the previous paragraph. Some other authors write that trust is "a mechanism for the reduction of social complexity" (book title of Luhmann, 1989) or that trust is a relation "TRUST(X, Y, C, τ, g_X)" (as already introduced in Section 2.4). In this case, the inner state trust is the output of the relation trust. So trust is a mechanism here that forms (or produces) the trust state. To distinguish both concepts of trust, I call the relation or mechanism the *trust mechanism* and the state the *trust attitude*. Together with the previous paragraph, the trust attitude can be represented as a trust value or a trust distribution.

Mutual awareness of the risk.   I do not require that the other one is aware that someone is trusting him. This may be the case and could make the compliance more likely. But it is no necessary feature for trusting. What is the reason for this view on trust? I consider trust to be the result of a cognitive process in the trusting agent. The fact that the trusted agent knows of the risk for the trusting agent can influence the trustee's behaviour and could therefore influence the degree of trust. But it is no condition that the underlying mechanism works. So this mechanism can form the trust attitude disregarding the mutual awareness of the trustor's risk. The result is some degree of trust.

In the case that the trusted agent does not know of the trustor's risk, it cannot act with goodwill towards the trusting agent. It needs not know that something is good or bad for the trustor. So goodwill cannot take effect in such a case. For this reason, the trusting agent may desire a loyal and benevolent action of the other agent, but this is no necessary criterion. The trust mechanism can work for every action of the trustee that is considered for evaluation.

Sources of trust.   Some definitions of trust include conditions trust depends on. For example, Castelfranchi and Falcone (2010) postulate that trust (as a mechanism) must depend on some trust sources like the ability and the willingness of the trustee. I omit such trust sources in the trust definition for two reasons. First, the influences of trust are still subject of research. It is hard to find new influences for a concept that is defined by its influences. Second, I consider the trust sources to depend on the implementation of the trust mechanism. Different application scenarios could require a different set of influences. The postulated conditions of trust usually refer to interpersonal trust. For me, this is just a specific implementation. To abstract from interpersonal trust, I leave the trust sources open.

With all these pre-considerations, the different forms of trust can be defined as follows.

Definition 2.1 (Probabilistic trust). Let P denote an agent and let a be a considered uncertain action of another agent O with the outcome $o(a)$. P's probabilistic trust is P's perceived certainty (subjective probability) about $o(a)$.

The outcome $o(a)$ can be understood as a random variable. Its possible values may indicate that a certain outcome will or will not be achieved (a boolean event); or the values may represent all possible outcomes. Associated with the random variable is a probabilistic trust distribution. It needs not be a probability distribution. It can, for example, be a belief function as well. A probabilistic trust value can be derived from the probabilistic trust distribution. So probabilistic trust can be understood as a probabilistic trust value or a probabilistic trust distribution.

The definition emphasises that trust handles the uncertainty that comes from the trustee's freedom to act (social complexity) – from his willingness and ability. If the trusting agent knew, how the other one decides (like a totally rational decision), and knew all input values of the decision process, then it also knew the decision and the action. So the other's action would not be uncertain in this case. No trust would take effect. With these considerations, the uncertainty requirement seems strong enough for me. Because of these arguments, I do not require a prisoner dilemma-like setting, in which the trusted agent gets higher payoffs when betraying. This setting just aims to ensure the uncertainty.

The definition does not regard the situational context explicitly. While it certainly influences the expectation, it is no necessary part of the definition. The situational context is already implied, because it affects the trustee's behaviour. And if the gain and loss of the trusting agent is regarded as below, it also influences the trusting agent's perceived problem.

Note that the definition does not speak of cooperation. It just mentions another agent's action. In this dissertation, I use the term cooperation to refer to all kinds of trust situations. The reason is that cooperation lets people think of a certain setting, which suits to understand trust. Just for this document, I consider all other kinds of trust situations as cooperation settings as well – which are degenerate though.

Definition 2.2 (Interest-related trust)**.** Let P be an agent that has some interest in the uncertain action a of another agent O, which leads to the outcome $o(a)$. P's interest-related trust is P's estimate of the chance and risk for P that is accompanied with $o(a)$.

As Jones (2002) notes, interest connects trust with risk (p. 230). So interest-related trust combines probabilistic trust with own gain and loss. Analogously to probabilistic trust, there can be an interest-related trust value and an interest-related trust distribution.

Definition 2.3 (Trust-related decision). A trust-related decision is a decision that depends on probabilistic or interest-related trust.

Influences that are not related to trust, may dominate the trust-related decision. The definition only says that trust (the reasoning about the other's behaviour) should have played a role. So a trust-related decision need not be a decision for a risk taking action.

Consider the following example: Paula could need Oliver's support to move to a new flat. She trusts Oliver to help her really well. But she also knows that Oliver is currently busy building his new house. So she decides not to ask him. This action of refused cooperation does not reflect her distrust in Oliver. It is simply influenced by other considerations than trust. The term trust-related decision should emphasise that trusting and deciding are different objects, although trust can impact a decision (see also Section 3.)

Definition 2.4 (Trust-related behaviour). Trust-related behaviour is behaviour that realises a trust-related decision.

Following this definition, trust-related behaviour includes to show no manifest action. Therefore it is no completely observable concept.

This dissertation focuses on the first step in the chain of trust concepts: probabilistic trust. All other forms of trust are derived from it. They should be subject of further research.

Further terms. The definitions above focus on the trust attitude. Then the trust mechanism is a mechanism that forms this attitude. Another interesting view centres the trust mechanism. It could be defined as a mechanism that models the behaviour of other agents. The behaviour of an agent could be regarded as its actions in response to certain situations. Then the trust mechanism can refine the model of the other's behaviour by observing the agent or talk with other about the agent, for example.

In a technical context, I call a specific implementation of the trust mechanism a *trust algorithm*. For example, the CREDIT algorithm of Ramchurn (2004) and the fuzzy approach of Falcone et al. (2003) are trust algorithms. A *trust model* is a conceptual representation of trust. It often refers to interpersonal trust. In this case, the model explains something real. But a trust model may also be a conceptual representation of technical forms of trust. Examples for this are the socio-cognitive trust model (Castelfranchi and Falcone, 2010) or the Enfident Model that is proposed later in this document. To regard anything like a trust model or a trust algorithm, I sometimes use the more general term *trust method*.

I consider *reputation* to be the view of a group or community about an agent's behaviour. This understanding of reputation is further detailed in Section 3.2.2. The view of a single third party on an agent's behaviour is sometimes regarded as a *recommendation*. Thus the fusion of the recommendations from a group of agents about the agent *O* could be understood as *O*'s reputation within that group. Other terms related to humans like *competence* or *acceptance* are described in Section 3.2.2. And some terms in the context of technical agents, like *ability*, *competence* and *willingness*, are given in Chapter 4. Note that a few terms are used for both, humans and technical systems. The context then determines the exact interpretation of them.

3 An Input-Output View on a Trusting Person

This chapter introduces interpersonal trust. Understanding this form of trust helps to understand trust between technical systems and to work across disciplines. Interpersonal trust is a working mechanism, so it can serve as a prototype. This chapter points at similarities and differences with a technical form of trust in order to increase the understanding about trust. Following Chapter 2, interpersonal trust is about the thinking, feeling, and behaving of a human being. Thus the matter of interest is what happens inside a human being. It is impossible to see that, though.

For this reason, I consider a human being a *black box*, a common approach in system theory (Ropohl, 1979; Trappl, 1983). Its inner configuration is unknown. An external entity can only observe what are the *inputs*, i.e. information and actions that affect the system, and what are the *outputs*, i.e. information and actions that the system produces. This consideration shows that a black box model has the inherent property of the separation between the outside (the environment) and the inside (the unknown configuration of the box) (Luhmann, 1985, Chap. 5). The system boundary in this chapter is the surface of the trusting person.

Why did I choose the paradigm of an input-output model to introduce interpersonal trust? One reason is that system theory is very suitable for interdisciplinary research (Ropohl, 1979; Trappl, 1983). Of course, the combination of trust and system theory is not new. For example, Luhmann (1979) investigated trust with social systems in mind. As a sociologist, he focused on the role of trust in the embedding of the individual in its social environment. This report distinguishes from others therein that it describes the trustor in a way that is pleasant for engineers: as an input-output system. The purpose of the model here is to easy the understanding of the trust construct. The model does not show the real constitution within an individual nor is it able to forecast the behaviour of an individual.

The black box description in this chapter is based on empirical studies of the literature. This approach – developing a model with support of empirical research – is good practise (Trappl, 1983), but meets a problem here. As Chapter 2 points out, the empirical research is often based on a different understanding of trust. Because of this, Narowski even devoted his complete doctoral dissertation (1974) to the analysis of the term trust and an attempt of operationalisation. He concludes, while he was able to integrate different views on trust in an overall definition of the construct, it was impossible to find clearly observable criteria for an operationalisation of the found construct. With this consideration in mind, the empirical results in this chapter must be seen as potential blocks for a building that reflects an empirically representation of trust, but not as hard facts.

Note that I model the trusting person only, although a model of the trustee would be very interesting as well. The trustee is omitted in the literature quite often. I make the same mistake here, as the model of a technical trustor is the common theme of this dissertation. The side of the trustee is subject to future research.

3.1 Overview

This section introduces the main components of the model and relates them to the literature. It includes the pure black box view as well as a rough concept of the internal structure. The internals of the box are just assumptions and, thus, are detailed only as far as they are necessary to understand the research on interpersonal trust. Sections 3.2 – 3.4 detail each component then.

3.1.1 Trust as a Dependent and Independent Variable

In psychology, it is common to look at a construct as a dependent and as an independent variable. When researchers consider trust as a dependent variable, they investigate how other variables constitute the variable trust. These other variables can be regarded as the "inputs of the trust mechanism". Gennerich (2000, pp. 12–21) identifies three groups of influencing variables:

- Trust as a variable of the trustor's personality,

- Trust as a variable of the trustee and

- Trust as a variable of the situation.

Following the input-output paradigm, the first group of properties lays inside the box, while both other groups are located outside the box. Only the second and the third group of influences are inputs of the black box. *Properties of the trustee* are the other's competence, empathy, loyalty or willingness, for example. The *situation* can be characterised by the time of the day or the fact that witnesses are present, for example.

All these properties strongly interact with each other as the following two examples illustrate. I trust my doctor to care about my health. But I will not let him repair my car. This example combines properties of the trustee and the situation. And when I drive with my car at night and I am stopped by the police, I could have a hard time feeling confident and establishing a trustful relationship with the police officer. This would be much easier if the same would happen during a nice sunny day. This example shows that the properties of the trustor's personality and the situation constitute trust only in their combination. So the inputs affect trust not independently like an additive model. Trust rather depends on the specific combination of the inputs together with the personality of the trustor.

Saying "Oliver is competent in repairing cars" seems to describe him. So "competent in repairing cars" is regarded as a property of Oliver. Another person would probably share the same view on him. "Oliver understands me" looks like a general property of Oliver as well. However this time, the speaker is part of the attribution as the object "me". Another speaker would possibly have a different view on Oliver. For this reason, the properties of the trustee are often regarded as the properties of the relationship to emphasise that they depend on both, the trustor and the trustee. Another reason is that human perception is very subjective; so the perception of the other one is always specific to the individual. Section 3.2 gives an overview of how the relationship and the situational context influence the constitution of trust.

When researchers consider trust as an independent variable, they assume a given state of trust and look, how it influences other variables like the communication in a group, conflict resolution, or personal satisfaction. Usually they want to show that high trust has a positive effect on the trusting person and the group or organisation the person belongs to. Only some researchers look at the *trust-related behaviour* directly (also called the *risk taking*). It may, for example, be a request for help or a self-explanatory statement. It is the immediate output of the black box. This action or verbal statement influences the environment of the trusting person and, in its consequence, leads to the mentioned *effects of trust*. Section 3.3 compiles selected investigations about trust-related behaviour and effects of trust.

The behaviour of the trustor transforms the environment. This, in turn, can result in modified inputs. So the trusting person is embedded in a kind of feedback loop. This consideration emphasises that, in the long term, trust is a property of the relationship between the trustor and the trustee, as both are in a continuous interplay.

Figure 3.1 illustrates that the trustor relates properties of the trustee and the situational context to trust-related behaviour. This relation is what Ropohl (1979, p. 55) calls the *functional conception* of the system. The glue between both sides lays inside the box and can thus not be observed. But from the input-output behaviour of the system, assumptions can be derived that help to describe this behaviour. There are many theories about how trust works in the human being. Most of them go beyond the intent of this chapter. Here they are only of interest as far as they help to understand the experimental research on interpersonal trust and to compare trust between humans with trust between technical systems in the succeeding chapters. The next section gives an conceptual overview of selected components in the black box.

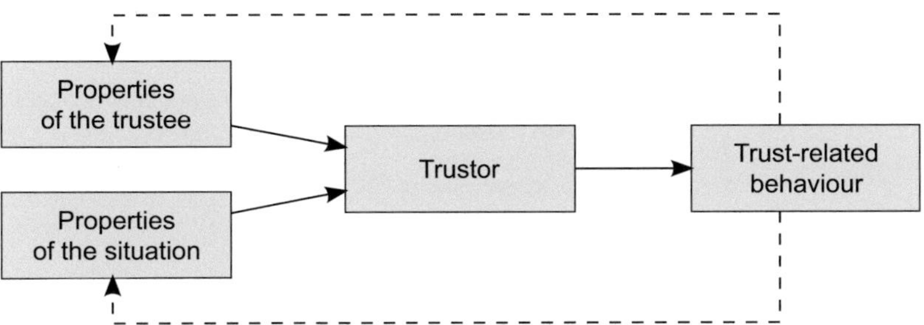

Figure 3.1: The main components considered when investigating the trusting person: Paula (the trustor) is in a situation, in which she considers to trust Oliver (the trustee). As a result of the consideration, she behaves somehow. Her behaviour will in turn influence the trustee and change the situation.

3.1.2 Attitude, Decision and Manifest Behaviour

In their article (1970), Kee and Knox review the research on trust at that time (mainly the work of Deutsch). They state that researchers had investigated trust with experimental games and operationalised their definition of trust accordingly. In their opinion, the observed behaviour in the game situations was subject to many interpretations though. Only some of the interpretation incorporated the concept of trust. As a consequence, they conceptualise trust in a generic way, in order that it is useful for different research methods, not just experimental games. Their main contribution is to conceive an inner *trust attitude* and separate it from the observable behaviour. The attitude is an internal state of the system, which makes it possible to consider the observable behaviour as well as answers in questionnaires and interviews to be indirect effects of the same inner state.

With the distinction of the trust attitude and the trust-related behaviour, the internals of a trusting person can be conceptualised as shown in Figure 3.2 on the next page. Paula *perceives* the situation and especially Oliver. This subjective process results in attitudes towards Oliver, like his perceived competence, consistency or acceptance. But it also constitutes *trust attitudes* towards the possible own choice in the trust situation – the trust-related be-

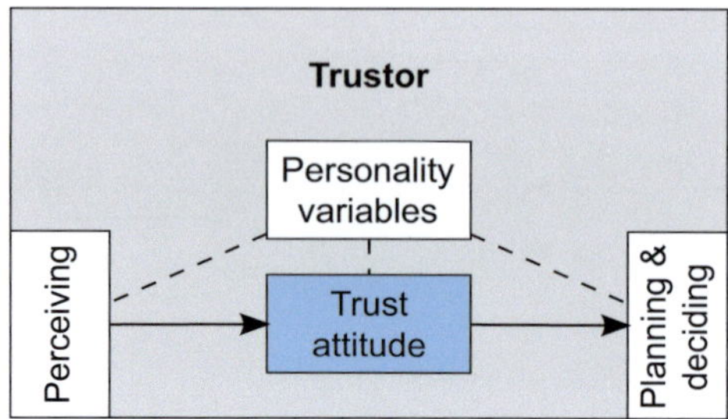

Figure 3.2: Concept of the internal structure in the trusting person: Paula (the trustor) perceives the external factors subjectively. From this subjective view on the trust situation, she constitutes trust attitudes regarding several possible actions. Finally she decides for one action, which becomes salient as her trust-related behaviour. All the processing is influenced by her personality and thus is subjective. This model is simplified and only focuses on selected trust-related aspects.

haviour. The attitudes represent expectations about the other's behaviour. They are internal states, feelings, and thus, a hypothetical concept.

The trust attitude is the inner opponent of the outside visible action: the trust-related behaviour. A component for *planning, decision making and control* transforms the trust attitude together with other influences into the finally executed actions. So the manifest behaviour is not a direct result of an individual's trust, but a consequence of various thoughts, evaluations and desires. Only some of them are related to trust. Section 2.5 shows this with the help of an example. Consequently it is impossible in general to conclude from a degree of trust directly to an action without considering further influences that are unrelated to trust.

Usually the trusting person cannot only choose between a completely trustful action and a completely distrusting action. Instead it can also opt for an action in between. For example, the set of choices, when closing a contract, could be to sign it as it is, not to sign it at all, to gather more information, or to insert an escape clause.

All the processing in the trusting person is influenced by her experiences and disposition: the perception, the trust development as well as the decision making (Forgas, 1985; Gennerich, 2000). These processes are very subjective. The additional state *personality variables* should indicate that the inner processes depend on the trustor's personality. This is only relevant from an observers point of view to understand the different behaviours of several trusting persons. The trustor himself needs not think about them. They are just there; the trustor cannot get around her subjective perception and judgement.

In summary, the blocks on the left and the right hand side of Figure 3.2 represent complex subsystems. The subsystem *perceiving* encapsulates all the processing from the sensors up to the interpretation of the perceived information. This process is very subjective. The subsystem *planning & deciding* transforms the trust attitude into behaviour. The trust attitude mediates between both. But it is not the only inner state of the trusting person that does so. The right block depends on many more attitudes, not just trust attitudes.

3.2 External Influences – the Inputs

The inputs are attributes of the environment that affect the investigated system. If the human body is considered a black box, the inputs are clearly associated with the sense organs: eyes, ears, nose, tongue, and skin. They capture input measures like light, sound, etc. These inputs do not suit to describe interpersonal interaction, though. A higher level of perception must be chosen.

In social psychology, person perception is concerned with complex properties of another person, like traits and motivation. Forgas (1985, p. 22) emphasises that this is very different from object perception, because the properties in person perception are hidden; they are subject to the interpretation of the perceiver. So interpersonal perception is to a lesser extend about pure perception but more about inferences. And both, direct perception and inferences, are very subjective in human beings.

The following section introduces concepts of interpersonal perception that are necessary to understand interpersonal effects in the development of trust. Then Section 3.2.2 details the influence of the other. *The subjective in-*

ferences about the other's behaviour develop trust. They are an appropriate description level for the system inputs. More specifically, they reflect the perception of the *relationship*, because they merge the perceivable properties of the other with the own personality. In the conceptualisation of Figure 3.2 on page 42, they are located at the output of the subsystem perceiving. The inferences at the output of this subsystem also suit to describe the influence of situational properties (Section 3.2.3). It is not part of this dissertation, though, what happens inside this subsystem. The end of this chapter discusses the interdependency between the inputs; they are mixed up when inferring.

3.2.1 Interpersonal Perception

Trust is directed towards another person; it depends on the relationship between the trustor and the trustee. How human beings perceive traits and form beliefs during the interaction with others, is investigated by the research area of *interpersonal perception* (German: *interpersonelle Wahrnehmung*). Its concepts help – as proposed by Gennerich (2000) – to better understand the interaction between the trustor and the trustee. This section introduces interpersonal perception as far as I use it in this report. It is mainly based on Fassheber et al., 1990, Chapter 4; Kenny, 1994, Chapter 1; Laing et al., 1966, Part 1; and Strack, 2004.

Paula's trust is based on her view on Oliver. She forms her opinion continuously during interaction. This process is subjective. During perception, her experience mediates her perception of Oliver's behaviour, resulting in the perceived behaviour. Thus two persons seeing the same situation, may perceive it differently.

For example, if a beggar asks another person for money, this person could think of him as a simple poor beggar or as a member of a Mafia-like beggar group, which is not poor at all. Besides situational influences, the view on the beggar depends on the personality of the perceiving person.

In interpersonal perception, researchers look at the interaction between all participants, not only at one person's view: Each person is "a self to himself each an other for the other, *together*, in relation" (Laing et al., 1966,

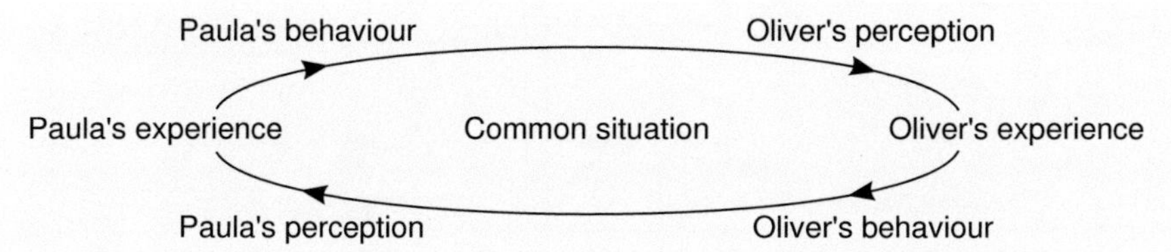

Figure 3.3: Interaction in a group of two persons (dyad). Interpersonal perception emphasises the existence of reciprocal effects. Compare this with Laing et al., 1966, p. 9.

p. 6). The interaction between Paula and Oliver can be modelled as a loop as drawn in Figure 3.3. Oliver subjectively perceives Paula's behaviour, which constitutes Oliver's experience, resulting in Oliver's decision and behaviour. Then the same happens vice versa on Paula's side. This mutual process is *interaction*.

So far, I introduced the basic idea of interpersonal perception. In the following, I distinguish several kinds of perception.

Paula may perceive Oliver as competent. This view of Paula on Oliver is called *other-perception* or a *perspective on the other* (German: *Fremdbild*): The one who perceives and the perceived person or object are different. I write this as $p(O)$, Paula's perspective on Oliver. Paula's perspective on a thing, opinion, relationship (X) is also other-perception written as $p(X)$. The perspectives $o(P)$ and $o(X)$ are other-perceptions of Oliver.

Oliver may see himself as clumsy. This time, the perceiver and the perceived person are the same. Such a view is called *self-perception* or a *perspective on myself* (German: *Selbstbild*). Paula's view on herself is also self-perception. Self-perception and other-perception form the group of *direct perspectives* (German: *direkte Perspektiven*).

In contrast to a direct perspective, a *meta-perspective* or *meta-perception* (German: *Meta-Perspektive*) is the perception about another perception. For example, Oliver may belief that Paula beliefs, he is going to cooperate. Such a belief of Oliver about Paula's belief about himself is written as $o(p(O))$

Figure 3.4: Meta-perception. This comic shows the development of a meta-meta-meta-meta-perspective ($p(o(p(o(p(X)))))$): "I see that he pretends not to know that I know that he knows that I would like to go for a walk."). With the kind permission of Bulls Press, © Solo Syndication/Distr. Bulls.

(German: *vermutetes Fremdbild*). The comic in Figure 3.4 shows different meta-perspectives.

Table 3.1 collects the perspectives used in this report. There a capital letter symbolises the object of perception. A small letter symbolises the viewing actor (e.g. p for Paula). Most perspectives look at the other one as a whole. The superscript *s* in the table indicates a situational view: How Paula currently perceives Oliver. Finally the superscript star means that this perspective refers to an ideal/desired conception. The ideal can be regarding competence, moral values, etc. I use my own notation derived from Fassheber et al. and Kenny but more convenient for engineers and consistent with other parts of this dissertation.

A perspective can also be the view of a group of persons or an institution. For example, the opinion of a trusted group or institution is important for reputation (see the paragraph about reputation in the following section).

According to Gennerich (2000, pp. 22–23), trust had mostly been investigated with the perspective $p(O)$ (e.g. by Rotter and Giffin) and correspondingly conceptualised as an attitude regarding O's personality. For him, self-perception $p(P)$ is also trust-relevant, especially in personal relationships.

Name of the perception	Short	Definition
Self-perception	$p(P)$	How does P perceive herself?
Perception of the other	$p(O)$	How does P perceive O?
Other-perception of myself	$o(P)$	How does O perceive P?
Ideal image of the other	$p^*(O)$	How does P wish O would be?
Situational perception of the other	$p^s(O)$	What behaviour of P does O currently perceive?
Supposed other-perception of myself	$p(o(P))$	How does P think O perceives her?
Supposed group perception of the other	$p(g(O))$	How does P think the group perceives O?

Table 3.1: Kinds of perceptions that are used to model trust in this report.

He concludes that *the trusting person evaluates relations of social percep-tions.* So trust develops depending on, how similar the other ($p(O)$) is to me ($p(P)$), or how the other relates myself ($p(o(P))$) to him ($p(O)$).

Summary. During interaction, an actor takes up different perspectives about herself/himself, about the other one or about things, opinions, relation-ships, etc. These perspectives in their combination determine the perceived relationship. Therefore they suit to systemically describe trust as a variable of the relationship. The next section does so.

Note. Fassheber et al. name the classification of the perspectives (within the research about interpersonal perception) *Sozialperspektivität* or *Sozia-le Perspektivität*. As a generic term for any kind of perspective, they use *Sozialperspektive* or *soziale Perspektive*.

3.2.2 Properties of the Relationship

Interpersonal trust is directed to another person (equivalent with alter, the trustee, or the other). In the second book of his *On Rhetoric*, Aristotle discusses that not only the arguments are important for persuasion, but also the way, how the listeners feel the speaker is disposed towards them. Aristotle names three properties of the speaker that make the audience trust him: "practical wisdom [phronēsis] and virtue [aretē] and good will [eunoia]" (Rhet. 1378a: Aristotle, 1991, p. 121). Butler (1991) calls these properties *conditions of trust* as they cause ("activate and sustain" (p. 644)) trust. He developed his *Conditions of Trust Inventory* in an elaborate study over several years. This inventory is a questionnaire to measure the causes of trust and mistrust in a specific person. In four sub-studies, Butler identified and validated nine factors: competence, discreetness, integrity, fairness & loyalty, openness, consistency, receptivity, availability, and promise fulfilment.

Gennerich (2000) collected the results of a few publications on this topic. In his opinion, the properties of the trustee cannot be separated from the interests of the trustor. Because of this consideration, he transformed them into *properties of the relationship* and developed a classification based on the concepts of interpersonal perception. Table 3.2 shows this classification. It groups properties of the trusted person depending on the perspectives they include. Then every combination of perspectives gets a group name assigned.

In this section, I mainly follow Gennerich's view and explanations while detailing each perspective with regard to trust. Note that Gennerich also discusses the perspective of the role model. If Paula perceives Oliver because of his role as an ideal for a certain aspect of her life, this is also a condition of trust. I omit it, as it is rarely found in the literature.

Perceived competence.   Trust refers to an expected behaviour, possibly related to a delegated task. This behaviour can require a specific competence of the trusted person. For example, to trust a doctor, his perceived medical competence is an important criterion.

Paula thinks Oliver is competent, if she perceives him ($p(O)$) as very similar to an ideal person ($p^*(O)$) for the given situation. This ideal person can represent a standard of performance respectively skills (expertise) or a standard of moral values (justice). When a leader represents a group, this ideal person is a prototype that incorporate all opinions and ideas within the group (German: Prototypikalität).

Competence: $p(O) - p^*(O)$	
Expertise:	O performs his tasks perfectly.
	O has the skills important for that task.
Justice:	O pays attention to laws and moral principles.
Prototype-ness:	O represents the group as a whole.
Consistency: $p(O) - p^s(O)$	
Predictability, dependability:	O behaves as he did before in similar situations.
	O behaves reliably and uniformly.
Honesty, integrity:	O speaks as he really thinks.
Promise fulfilment:	O does what he promises.
Discreetness:	O can keep confidences.
Similarity: $p(P) - p(O)$	
	O has the same goals or moral values as I have.
Empathy: $p(o(P)) - p(P)$	
Understanding:	O sees me as I am.
Receptivity:	O understands my ideas.
Acceptance: $p(o(P)) - p(O)$	
	O takes me as I am and stands by me.
Benevolence:	O can relate myself positively to himself.
Loyalty:	O protects me and makes me look good.
Reputation: $p(g(O)) - p(O)$ with regard to $p^*(O)$	
	Others compliment on O's work.
Openness: $p(o(P)) - p(O)$ and $p(O) - p^s(O)$	
	O tells me his ideas.
	(acceptance and honesty)
Respectfulness: $p(o(P)) - p(P)$ and $p(o(P)) - p(O)$	
	O incorporates my view in his behaviour.
	(understanding and acceptance)
Availability: $p(o(P)) - p(O)$ and $p(O) - p(P)$	
	O is present when I need him.
	(acceptance and closeness)

Table 3.2: Trust-related properties of the trustee. These properties are classified and expressed by relations of perspectives. The lower three properties are dependent in that they are combinations of the upper six. (Adapted from Gennerich, 2000, p. 24)

Butler (1991) investigated competence in terms of skills. (It is named expertise in Table 3.2). He considers competence a very important property of the trustee (p. 646). Butler also identifies the property fairness, which is similar to justice. However he uses it more like benevolence or loyalty (see the paragraph about perceived acceptance below).

Perceived consistency. Luhmann (1979) describes trust as a commitment to one's own expectation about the future. However Oliver must have shown consistent behaviour in the past, so that Paula can predict his behaviour. The prediction results in the expectation, which in turn is evaluated with regard to trust.

Oliver's behaviour appears consistent for Paula, if this current behaviour ($p^s(O)$) can be predicted to his past behaviour, which can be considered to be part of Oliver's personality ($p(O)$). Similarly Paula perceives honesty, if Oliver's current statement ($p^s(O)$) is consistent with his supposed state of knowledge ($p(O)$). And finally, Oliver's former promise ($p(O)$) is fulfilled in his current actions ($p^s(O)$). Discreetness is a special kind of promise fulfilment.

Butler identified four factors belonging to this group: discreetness, integrity, consistency, and promise fulfilment. He describes integrity as "honesty and truthfulness" and consistency as "reliability, predictability, and good judgement" (pp. 646–647). Butler also measured overall trust in a specific person. The items of overall trust mainly loaded on the factor integrity. They also loaded on the factors fairness & loyalty as well as discreetness, but more weakly. Thus the attribution of integrity is the most important trait to judge the trustworthiness of the other.

Perceived similarity. Sitkin and Roth (1993) investigated similarity with regard to values. When Paula perceives that Oliver has different cultural values than her, she will doubt his view of the world. "The threat of future violations of expectations arises because the person is now seen as a cultural outsider—as one who 'doesn't think like us' and may, therefore, do the 'unthinkable'" (p. 371). Thus the value incongruence could compromise Paula's concerns. Gennerich takes a step forward. The value incongruence could let Oliver condemn her and so compromise her self-conception.

Gennerich also discusses reports about attraction to differences. If the other definitely accepts one's own person, similarity is not necessary any more. In this respect, similarity should only ensure acceptance in the beginning, as long as the acceptance of differences is not sure.

Also not that similarity can conflict with competence, although both are conditions of trust. For a professional and distant relationship, competence might be more important, as Section 3.4 shows.

Perceived empathy. According to Kee and Knox (1970), both, Paula and Oliver, must be aware of her risk in the trust situation. To trust Oliver, Paula must be confident that he knows about her situation and her act of trusting.

This knowledge is only a weak form of empathy though. It does not reflect the importance of perceived understanding for trust development. If Paula thinks Oliver perceives her ($p(o(P))$) differently than she perceives herself ($p(P)$), she will accuse Oliver of that. This difference compromises her own identity, because "self-identity is a synthesis of my looking at me with my view of others' view of me" (Laing et al., 1966, p. 5). Gennerich infers that the self-identity is – besides others – a precious property Paula risks in her act of trusting. Being understood supports the own identity and gives a feeling of safeness, necessary to trust.

Butler calls one of his factors receptivity. He means with it whether Oliver is open for Paula's ideas. This is a form of understanding.

Perceived acceptance. To perceive acceptance, Paula must see that Oliver can relate her ($p(o(P))$) to himself ($p(O)$). Oliver must be able to integrate her personality in his life. So acceptance goes beyond simple similarity, in that it allows both parties to be different. If Oliver can accept the strange, the different sides of Paula, she can trust him despite the risk that comes from the personal differences. "Acceptance can be considered a form of behaviour here, which makes it possible that the dissimilarity of two persons does not violate their identities (Gennerich, 2000, p. 46).

Butler identified a factor which is loaded by items about loyalty and fairness. In his opinion, these two properties "are conceptually similar in that they both refer to one's perception of another's concern for one's welfare. However, fairness refers to perceived equity; loyalty, to perceived benevolence" (p. 652). With this definition, he focuses on the favourable aspect of fairness in contrast to justice with its more professional, distant meaning.

This relation of perspectives covers trust reciprocity, too. Especially in personal relations, it may be important for trusting him that the other trusts me (Butler, 1986). This is a form of acceptance.

Perceived reputation.   Reputation is the opinion of others about a person or community. In the development of trust, it is mainly used to validate one's own opinion. As such, the relation between both, the own opinion ($p(O)$) and the opinion of a group ($a(g(O))$), is evaluated with regard to an ideal type ($p^*(O)$). As a consequence, the third-party information can exceed one's own experience in such a way that it even destroys trust (Gennerich, 2000, pp. 30–31).

Besides the verification purpose, reputation can create a kind of impersonal trust, if one's own experience with the other is too small (Gennerich, 2000, p. 30). This process can be modelled with Coleman's (2000, Chap. 7) *principal-agent* schema. The principal delegates the acquisition of trustworthiness to the agent, who in turn determines alter's competence or consistency. Strasser and Voswinkel (1997) mention certificates as one form of realisation. Today many parts of professional life depend on recommendation letters of experts, school reports, academic or organisational titles, seals of quality, etc. Certificates can even create trust, if alter is not present. In some of the mentioned examples, the agent is an institution. Strasser and Voswinkel call such an agent a *trust agency* (German: *Vertrauensagentur*). Examples of trust agencies are product testing institutions (Stiftung Warentest or Consumer Reports), credit protection agencies (Experian or Schufa), and literary critics. Strasser and Voswinkel also note that trust agencies must be suspicion agencies at the same time. This is, because the principal only perceives them as trustworthy, if they discover and punish black sheep (agents who betray).

Besides the principal and the agent, alter is also concerned with his reputation. Companies care for their corporate identity and, indeed, try to design their publicly perceived image (Strasser and Voswinkel, 1997). The trust in an organisation manifests in the trust towards its representatives and the other way around. One trusts those representatives because of their (known) social identity, not because of their (unknown) character. Strasser and Voswinkel (1997, p. 225) call the trust in a specific social identity of a person *role trust* (German: *Rollenvertrauen*), Gennerich (2000, p. 40) *transpersonal trust* (German: *Transpersonales Vertrauen*).

The described formalised reputation process helps to establish trust between frequently changing and foreign actors. This is typical for modern societies with a wide division of labour. Sometimes, reputation is even established based on a chain of contracts or certificates (Strasser and Voswinkel, 1997).

The theoretical concepts introduced above are only partly accompanied by empirical research. Empirical studies mainly investigated reputation through the structure of the social network.

Burt and Knez (1995) analysed the data of a survey among senior managers about their network structure and success. The analysis showed for a strong relationship that trust in a specific other person is higher, if the relationship is embedded in a joint dense network, that is, if both have many third parties in common. On the other hand side in a weak relationship, distrust in a specific other person is higher, if the relationship is isolated from the network, that is, if both have only few third parties in common. The study analysis only one point in time; as such, it cannot draw the underlying dynamics in form of a cause-and-effect chain. Burt and Knez suppose that the gossip of third parties mainly validates and thus confirms one's own opinion.

Buskens (1998) modelled and simulated the influence of the network structure on the trust level in a buyer-seller-scenario. The network structure is characterised by the ties between the buyers, which represent the information flow. When looking at one individual buyer, a higher number of outgoing information connections to other buyers (higher centrality) increases the situational trust level of that individual buyer, because he has a larger reputation effect and so is more dangerous for the seller. When looking at the network globally, a higher density of network ties increases the mean trust level. Some other results were unintuitive, though.

Both studies above discuss the effect of reputation on the development of trust towards a specific person. Reputation mainly serves to verify one's own opinion about the other. The verification often works as a confirmation of trust or distrust. All in all, there is still need for empirical investigations into the effect of reputation on trust (Burt and Knez, 1995).

Perceived openness, respectfulness, and availability. The previous explanations have shown that a perspective represents not only one trait of a person, but a group of adjunctive traits. In turn, a trait names a specific aspect of a perspective. It is likely that the perspectives have more aspects as listed in Table 3.2. This section shows that such aspects must be combined to describe some other traits.

When Paula feels Oliver is an open person, she perceives that he integrates her in his life (an aspect of acceptance) and tells what he thinks. Thus, openness follows from honesty and acceptance. In the same way, availability combines aspects of acceptance and similarity (geographical closeness), while respectfulness combines aspects of empathy and acceptance. These perspectives depend on other perspectives. For this reason, I separated them from the independent perspectives in Table 3.2.

Butler investigated openness and availability. Gennerich lists a few authors for all three of those combined perspectives.

3.2.3 Properties of the Situation

So far, I showed that trust depends on properties of the relationship, not just on those of the other one. In the model introduced in Section 3.1, trust also depends on variables characterising the situation. Many researchers have investigated this situational influence empirically. In this section, I give an impression on the topic by introducing a selection of these studies.

The situation sets the need for a decision, to trust or not to trust, because it has an inherent complexity associated with risk. Thus one group of situational variables determines the perceived complexity and risk. These variables directly affect the perception of the trust situation. The variables in the second group modify, how the trusting person perceives the properties of the relationship as described in the previous section. The trustor takes these modified properties into account when deciding to trust or not to trust. So such situational characteristics affect the trust development only indirectly.

Situational properties affecting the perceived risk.   Several situational methods arose in the past, to strengthen trust for a specific decision: a ceremonial oath, a testimony in presence of a witness, depositing a pledge, etc. They all reduce the perceived risk of the situation (Strasser and Voswinkel, 1997). Heisig (1997) argues in the same direction when he says that the chance of trust betrayal is low even without formal sanctions because of social forces. The fear to be stigmatised in one's social environment, for example, acts deterrently.

A very interesting study of Oswald (1997) explored, whether an increasing complexity of the situation increases the trust in a consultant. The subject, in the role of a mayor, had to accept or reject an investor with the help of a consultant. In the experiment, three different ways to explain the economic connections were used, resulting in a simple, a medium complex, and a very complex situation for the subject. In addition, either an appropriate or an inappropriately simple consulting concept was offered.

It appeared that the subject perceived an increasing competence of the consultant with an increasing complexity of the situation. The perceived sympathy remained equal. Both together let to an increasing trust in the consultant with an increasing complexity. This applies for both consulting concepts. In addition, the subjects increasingly accepted the consulting concepts, especially the inappropriately simple one. Oswald concludes that the need for trust had increased because of the complexity. This need had resulted in the changed perception. All in all, this study shows that the perceived complexity of the situation influences the trust in the trusted person.

Situational properties affecting the perception of the trustee. Communication is an important vehicle to perceive properties of the relationship. In a game situation, an increasing amount of communication resulted in an increasing amount of perceived mutual trust (Loomis, 1959). In a real life situation, the perceived relationship does not only depend on the amount of communication but also on its content (Alexander et al., 1989). Trust also improves the communication in an organisation. This is important for efficient work processes (Bierhoff, 1995). Note on all these investigations that trust and communication influence each other strongly. Hence it can hardly be separated, what comes first and what is the effect of the other (Kassebaum, 2004, p. 49).

The trusting person could also communicate with a third party. Then the communication forms the reputation of alter as a situational property (Gennerich, 2000, p. 18). Reputation is discussed as social perception in Section 3.2.2.

In literature, there are some indications that power, control, and punishment hinder the development of trust:

Strickland (1958) investigated a surveillance game situation, in which the subject played a supervisor who could monitor the result of her/his subordinates. However these results indeed were fixed, because the subordinates were only faked. Thus trust development was based only on the supervisor's own attribution – without any evidence. In summary, the monitoring activity of the supervisor led to lower trust in the monitored subordinate. Strickland argues that trust development needs opportunities to be disloyal. Monitoring avoids these opportunities.

A study of Tedeschi et al. (1969) shows the influence of social power on the trust development. In a game situation with equal social power, subjects cooperated more often in reaction on a message with an intent to cooperate than in situations with unequal social power. In addition, their own messages were true more often. In a situation with a decline in power, weak and strong subjects acted quite competitive to maximise their gains.

3.3 External Effects – the Outputs

Outputs are attributes of the system that affect the environment. A human being affects the environment through verbal and non-verbal actions. In this section, such behaviour is of interest, if it may be driven by trust. In the following, the immediate behaviour of the trusting person is distinguished from its secondary effects. The behaviour is the real output of the system. The effects rather help to see the advantage of trust.

Trustful behaviour.   In the literature, the forms of real-life trustful behaviour are only investigated rarely. Most often, the authors were interested in positive effects of trust. Petermann (1996) is one of the few who study trust with behavioural monitoring – making the immediate trust-driven behaviour of the subjects the first-class object of the investigation. He classifies trustful behaviour according to four features:

Self-exploratory statements open the own personality to the other. They disclose information the other could misuse. Paula could fear that others – if they come to know the information from Oliver – could misunderstand it in a way that they think negative about her. In this case, her standing in the group, that is, her social identity is risked. Thus trust is necessary for such statements.

Here-and-now statements about the current situation involve risk because the reaction of the other cannot be known in advance. This is in contrast to well-proved statements about previous experience, general knowledge, or negligible information. This uncertainty requires trust.

A request for help gives the control to the other, on whom the requesting person then depend. This involves trust. Often a request for help also admits the own weakness and vulnerability. Such a request can be verbal or non-verbal. For example, signing a contract is a non-verbal form of agreement about help.

A request for feedback can be understood as a request for help. However
it differs in the way that it allows a negative reaction of the other. The
requesting person makes herself even more vulnerable.

All items involve uncertainty and the risk to get disliked, exploited, dis-
appointed, or punished. They rely on trust. This list does not claim to be
complete, though. Petermann selected those items that are applicable to a
dyad and integrable in an interaction sequence. He gives further references
for other views.

Effects of trust.   Trustful behaviour, that is the direct output of the system,
results in indirect, mostly positive effects. They underline the importance of
trust and may serve as motivation. Because of that, I give a rough overview
in the following, although the effects of trust are not directly related to the
system description. I follow Rotter's (1980) distinction between personal
consequences and consequences for the society – or in general, for the
affected community.

Personal consequences include that "the high truster is less likely to be
unhappy, conflicted, or maladjusted; he or she is liked more and is sought
out as a friend more often, both by low-trusting and by high-trusting others"
(Rotter, 1980, p. 6). The high trustor opens up to others more easily and can
better interpret non-verbal behaviour of others; nevertheless he is not more
gullible (Koller, 1997, p. 16). In addition, trust lowers stress in relationships
(Gennerich, 2000, p. 11). This might be related to Oswald's (1997) report
that trust is important *to adapt the cognitive resources to the complexity of
the situation.*

There are also advantages for the communities the trustor belongs to.
"People who trust more are less likely to lie and are possibly less likely to
cheat or steal. They are more likely to give others a second change and to
respect the rights of others" (Rotter, 1980, p. 6). Gennerich (2000, p. 12)
compiled studies from the area of market and organisation and concluded
that trust has positive consequences on the quality and quantity of the information flow, on the acceptance of information, on cooperation and negotiation success, on the engagement within the organisation, on the general

and team performance, and on the satisfaction. *All in all, trust is important for successful professional collaboration.*

3.4 Inner Processing

The previous sections mainly dealt with the observable environment of the black box in form of related studies. (Describing the interaction partner with the concepts of person perception left the observable level, of course.) This section presents assumptions about what happens in the human being. There are many theories on the subject. I only present three selected topics, which help to relate interpersonal trust with trust between technical systems. They are regarded in the succeeding chapters.

3.4.1 Trust as a Generalised and a Specific Expectation

In unknown situations, no experiences are available to support the decision. Human beings have some amount of trust to cope with such situations. This kind of trust is developed as a basic sense in the childhood (*basic trust*, German: *Urvertrauen*) and is continuously learned by generalising over previous experiences later. It is a *generalised expectancy* about the statements and actions of unknown others (Erikson, 1968; McKnight and Chervany, 2001; Rotter, 1980).

This kind of trust can be measured with questionnaires as a one-dimensional magnitude (e.g. Couch and Jones, 1997; Kassebaum, 2004). Persons who have high values on this component and those who have low values are called *high trustors* and *low trustors*, respectively (German: *Personen mit geringer/hoher Vertrauensbereitschaft*). Rotter (1980) thinks that high trustors grant a credit to their interaction partners if no clear-cut data is available. This credit lasts until clear evidence makes it impossible to trust. With this strategy, high trustors can better exploit chances. This is different to gullible or naive people, who trust, even if there is clear evidence not to trust. On the other hand, low trustors start with suspicion in the absence of clear-cut data. Trust must develop in a process of proving. With this strategy, low trustors take less risk.

The following two experiments illustrate the differing strategies: In a game experiment (Parks et al., 1996), high trustors reacted with increased cooperation on a message that the unknown other wants to cooperate, low trustors did not. In contrast, low trustors decreased cooperation after a message of the other that indicated competition, high trustors did not. In a perception experiment (Gurtman and Lion, 1982), low trustors recognised connotatively negative words faster than other words, while high trustors recognised all words in about the same time. So the perception of the other and the situation depends on the personality of the trustor. Low trustors seem to be more sensible about risk.

This unspecific expectancy is necessary at the beginning of an acquaintance when there is insufficient information about the other one. In the course of the aquaintance, direct experiences with the other one provide the necessary information to form a *specific expectancy* (McKnight and Chervany, 2001). Trust then depends on the perceived relationship. This form of relation-specific trust can also be measured with questionnaires. They show that high and low trustors can develop strong trust in a specific person likewise (Couch and Jones, 1997; Kassebaum, 2004).

The trust in a specific other is still a generalisation over several situations, but regards a fixed person. The trust in a single trust situation is the most specific judgement. So while trust as a generalised expectancy is considered a stable trait of some one, the trust in a specific situation depends on the properties of that situation and the experience with similar situations (McKnight and Chervany, 2001).

What does all that tell us for this dissertation? First, a trust mechanism needs the ability to derive trust from past experiences with similar situations: it generalises over the similar experiences. Note that here, the situation includes the trustee. If the current situation is rather new and unknown, all existing experiences are taken. This results in a rather stable generalised expectancy for all unknown situations. The more the properties of the current situations are similar to those of experiences (including the properties of the trustee), the more specific the trust becomes and the more it may differ

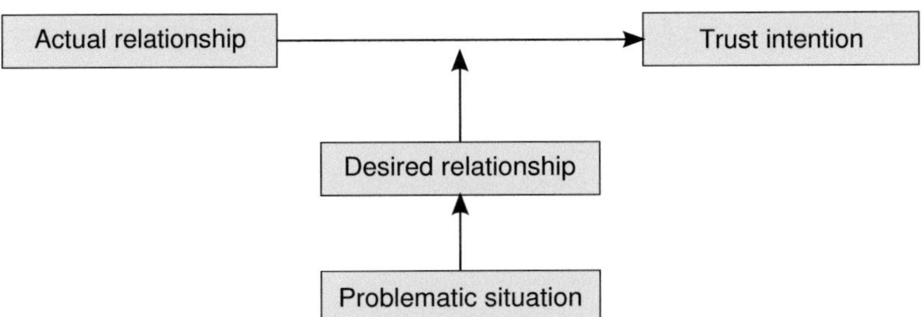

Figure 3.5: Model of trust development according to Gennerich (2000, p. 62). In a given situation, trust is high if the relationship with the trusted person matches the need.

from the generalised expectancy. This way, the trust mechanism specialises on the trustee and the situational context. Or in other word: To judge the current situation, the trust mechanisms takes all experiences, while those that are more similar to the current situation have a stronger influence on the resulting trust than those that are less similar. This is postulated as a requirement in Section 4.4.2.

Second, a person can form a trust attitude towards various classes of situations, not only towards a completely specific situation. Someone can have a trust expectation about a specific person, about a completely unknown situation, about his family and friends together (Couch and Jones, 1997), about a certain company (McKnight and Chervany, 2001), etc. If some properties interest are independent from each other and each property defines a class of trust situations, then independent trust components could be found for each class. This feature of interpersonal trust let to the concept of querying for the Enfident Model as introduced in Section 6.4. Querying describes that the trust mechanism can reason about the current situation but can potentially infer many other magnitudes too. The querying mechanism is only conceptual though. A trust algorithm needs not implement it.

3.4.2 Matching of the Actual and the Desired Relationship

Gennerich (2000) takes a different approach to describe how trust is formed

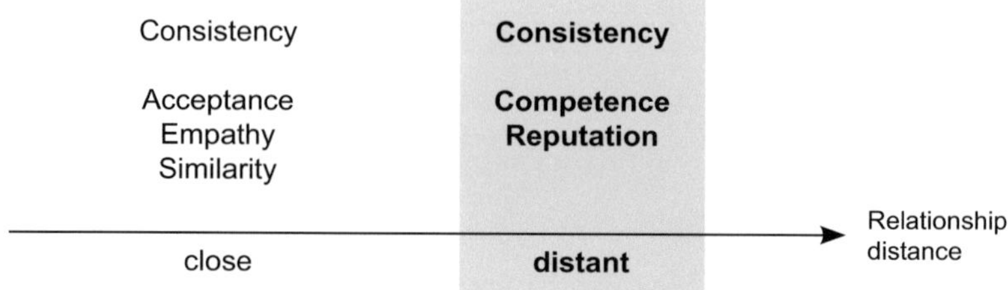

Figure 3.6: Importance of the trustee's properties depending on the distance of the relationship according to Gennerich (2000).

in a given situation. It is illustrated in Figure 3.5 on the previous page. For him, the situation defines a problem for the individual. To solve the problem, another person could be helpful. This results in a desired relationship, which is not connected with a certain other. It just describes what kind of relationship would be ideal to solve the problem. Then the relationship with a specific other person can be compared with this ideal. The closer both are, the higher is the trust intention. Note that the trust intention is the inner trust state in Gennerich's text.

Thus the model combines the situational context with the trusted person to form trust. This is like others do. The properties of the trusting person are hidden, on the one hand, in the word relationship and, on the other hand, in the way, how the situation is perceived as a problem and transformed in a desired relationship. The model emphasises that there is a problem to address with the help of another person. The trustworthiness of this other person depends on that problem.

The problem, in turn, is defined by the situation. When comparing this with the situational influences described in Section 3.2.3, the first group of influences that define the perceived complexity and risk, are clearly assigned to the problem and result in a certain desired relationship. In contrast, the second group of influences that affect the perception of the trustee, may result in a certain desired relationship but may also be perceived as a property of the present actual relationship.

A first class problem for humans is identity building. A personal identity is supported by relationships with similarity, empathy and acceptance. For example, Paula could talk with Oliver about her problems with her mother. She might want to get support for her position (stabilise her personal identity), but also some hints how to handle the situation in a better way (solve a personal problem).

In contrast, a professional context needs more competence and reputation. When Paula brings her car to a garage, for example, she wants to know what is defect with her car (help with a professional subject) and to be treated as a respected customer (meet her social identity), while she might have talked with the mechanic Oliver never before.

Gennerich distinguishes close and distant relationships and connects them with the perspectives of interpersonal perception that were introduced in Section 3.2.2. Figure 3.6 visualises this assignment and the above scenarios are examples for a close and a distant relationship, respectively. The assignment of perspectives to the one or the other kind of relationship is weak. It should express that the given perspective is especially important for the assigned context. Consistency is essential for all kinds of relationship. Without a consistent behaviour, no clear expectation could be formed.

Personally, I agree with the model of Gennerich, because it feels like a good tool to analyse trust situations. It shows that trust is not a rating of the relationship but the matching of a problem with the relationship. The problem requires a certain kind of relationship, that is, a certain way the trustor wants to be treated. The model of Gennerich helps to analyse trust situations between persons accordingly.

But how does the model relate to trust between technical systems? When technical systems interact, they do so to fulfil a certain job. They focus on a professional subject matter. In Gennerich's classification, this is a distant relationship. Consequently for trust between technical system, consistency, competence and reputation are mainly effective. So it is sufficient to focus on that kind of relationship in this dissertation.

In addition, the idea of matching a problem with a solution is quite familiar to engineers. This view helps to understand trust.

On the other hand, the model shows how trust between persons and trust between technical systems differ. Close relationships to not matter in the technical case, because machines do not have something like an identity that needs to be formed and maintained. This is a major difference. It makes clear that results from interpersonal trust should only be transferred to technical systems with care.

3.4.3 Self-Perception

Finally, I want to point to two studies of (Koller, 1988) and Zak et al. (1998). They investigated independently of each other whether interpersonal trust develops in response to one's own actions. Following self-perception theory (Bem, 1972), people develop their attitude by observing the own action and inferring what attitude it comes from.

In the experiment, every individual had to perform an action that requires trust in another person. Directly before this action, the trust is the other person is measured with a questionnaire. This is also done directly after the trustful action and before the trusted person could comply and betray. So the amount of positive or negative experiences is not changed during the experiment. The only thing that really happened before the two trust measurements is the trustful action of the trustee. It turned out that trust increased in the consequence of the own action and the amount of increase was related to the risk the trustful action involved.

The experiment shows that the own action caused the increase in trust, not the behaviour of the trustee. This sounds unexpected, because the purposes of trust as introduced in Section 2.2 indicate the trust is a mechanism to handle the uncertainty that comes along with the trustee.

Koller (1997, 1988) argues that the decision to take a risk in a trust situation depends not only on the amount of positive experience with the interaction partner, but also on the importance of the goal that can only be reached with the help of the other one. A trust situation comes along with a lack on control. Trust creates an illusion of control by overestimating the competence and reliability of the trustee. This illusion transforms the feeling of helplessness in a feeling of confidence, so that the goal could finally be reached. These considerations could also explain the findings of Oswald (1997).

With regard to technical systems, it seems desirable that trust should depend on the performance of the trustee, not just on own actions without any exchange. People expect that a machine is a somewhat rational device. In summary, this section should make clear that there might be major differences in the working of interpersonal trust and trust between technical systems.

3.5 Summary

This chapter introduced interpersonal trust in order to deepen the understanding of trust in general as well as to highlight similarities and differences between interpersonal trust and inter-machine trust. This way, human beings can serve as a prototype for some aspects of trust between technical systems. To relate both disciplines with each other, I found it useful to consider the trusting person with regard to her trust-related processing as an input-output system. This is a view often taken in technical fields.

An input-output model clarifies what is outside and what is inside the system. The objects outside the system can be observed by an investigator, while the processing inside the system can only be assumed; it is subject to theories.

Trust is conceptualised as an inner state of the system. It is not directly observable, but effects trust-related actions and statements (including those in questionnaires), also called risk taking actions. These are the observable system outputs. They are rarely investigated though. Usually secondary effects like improved communication or happiness are the subject of research.

The expectation of the trusting person in a specific situation, that is the cooperation outcome, depends clearly on the person, whom is trusted, and on the situational context. So it is plausible that researchers found them to be the inputs of the system.

To describe the influence of the interaction partner, directly observable properties are inappropriate. The perceiving trustor implicitly makes inferences about the other's traits and motives. In the model, this happens in the perception component, a conceptual sub-system that precedes the trust component. The traits and motives are the outputs of this sub-system. So the concepts of person perception suit better for this purpose. Relations of perspectives as perceived by the trusting person allow a conclusive description of the inputs. This idea has been proposed by Gennerich (2000). It is borrowed for this dissertation, because it is very systematic and general. Understanding it may help to design a trust algorithm for a specific application.

Gennerich identifies seven perceived relations: Consistency (including predictability and honesty), competence (including expertise and justice), similarity, empathy, acceptance (including loyalty and benevolence), reputation and role model. They are all used to form the expectation about the cooperation outcome. The reader may wonder why just predictability is not sufficient for this. If the behaviour of the other one is predictable, that would be sufficient to form a clear expectation; and if it is unpredictable, no expectation for a single outcome could be found.

Predictability describes the other's behaviour over several interactions in similar situations. It applies to all kinds of trust situations. In contrast, most other characteristics like expertise, honesty, empathy or acceptance reflect abilities that are necessary to solve the one or the other problem. The trusting person may need some one to repair her car, to talk about her mother, to talk about religion or to share a secret. Different abilities are necessary for these cooperation tasks, but all of them can be realised in a predictable way. Therefore to form a trust expectation, the individual must judge whether the cooperation partner has the necessary abilities and whether he applies his abilities in a predictable way. Falcone et al. (2003) identify in addition that the other one must be willing to perform the task. So the predictability describes whether the trusting person can well model the other's abilities and willingness. If the model is insufficient, it results in wrong expectations and thus unpredicted behaviour.

The studies about the characteristics of the interaction partner present correlations between those characteristics and the degree of trust. The correlations can be explained in the above way that trust is a magnitude that is derived from those characteristics. Or they can be explained in the way that the trust attitude and the other attitudes towards the relationship with the trustee depend all on common subsets of other internal states; they have common causes. These joint sources may indeed exist, because all these attitudes are mainly based on previous interactions with this opponent. So these interactions are the experience that drives the formation of attitudes towards the trusted person. Both points of view are valid, because they cannot be distinguished by an external observer. Section 10.2.2 further discusses the subject with regard to the Enfident Model.

The situational context defines the problem for the trusting person. For this problem a good solution with a good result is desired. The individual can match the problem with the interaction partner. So both together form the trust expectation. On the other hand, the trusted person faces this situation as well. He first agrees to do the task that is inherent in the problem (if there is an explicit agreement) and later actually performs it. So the situational context and the intended interaction partner are in a relationship from both, the trustor's and the trustee's point of view.

The inputs and the outputs are the observable part of the model, the functional conception of the system. This dissertation does not cover decision and action. Instead it stops when the trust expectation is obtained. So for this text, the inputs are more interesting than the outputs of the black box. The inputs of the Enfident Model are compiled in Section 6.2.

The set up and processing within the human being can only be assumed. The presented model conceptualises three main components that are related to trust. The perception component observes the environment and makes inferences about it. It works subjectively and, thus, reflects, what the individual thinks, not what is true. The trust component forms the trust attitude towards the present trust situation, which is an inner state of the system. To do this, it takes the outputs of the perception component. This component also works subjectively. For example, some people are more suspicious than others. The trust attitude than influences the planning and decision making component. Other considerations and attitudes affect this process as well. So the resulting decision and trust-related behaviour may reflect the trust attitude but may as well be quite different from it. If this set up is applied to technical systems, this dissertation deals with the trust component only.

The trust component learns for a specific situational context and interaction partner. But it is also able to generalise over similar contexts and partners in order to judge new settings. This functionality is discussed for the Enfident Model in Section 6.3. The model of Gennerich (2000) emphasises that the trusting person needs to match the situational problem with the interaction partner, which should help with the problem. As a consequence, different characteristics of the interaction partner influence the trust attitude depending on the kind of situation. So trust can also be considered as a problem-related evaluation of the trusted person. Finally some studies suggest that trust is developed not only in response to the other's behaviour but also to own actions.

In summary, looking at interpersonal trust gives some interesting insides and helps to understand trust. In the consequence, some ideas of the Enfident Model are inspired by the thoughts in this chapter.

4 Requirements and Related Work

This chapter reflect on features of the mechanism trust and relates them to the work of others. These features are requirements for a new trust algorithm. In some cases, the introduction of the literature leads to a feature in the text. In other cases, own pre-considerations formulate a feature, while references are given afterwards. For the literature review, I chose only few trust algorithms, although a vast number of papers could be found. Two aims guided the selection: First, the review should present a wide range of trust concepts to explain the various features of trust. Second, the papers should cover a wide range of applications. This should emphasise the general utility of trust and the general applicability of the required features.

In data networks, trust helps to rate the reliability of the exchanged content. The selected literature includes methods for wireless sensor networks (Zhang et al., 2006), vehicular networks (Bamberger et al., 2010, 2012; Golle et al., 2004), and general ephemeral networks (Raya et al., 2008). The trust literature in the field of multi-agent systems usually considers the scenario of trading agents at a virtual market place. Trust should arise between buyers and sellers (Rettinger et al., 2008; Ramchurn, 2004; Sabater and Sierra, 2002; ElSalamouny et al., 2010).

The work of Falcone et al. (2003) is conceptual. They combine beliefs to obtain a trust belief. How they obtain the first level of beliefs remains open though. It seems they want to model human trust with a computational formalism. The dissertation of Marsh (1994) is conceptual in a similar way.

Section 1.1 already clarified that this dissertation focuses on the trust development in the individual only. Trust-related subjects on the society level are omitted, although they are certainly important for a trust-based society of technical systems. They include, for example: norms and policies (Ramchurn, 2004; Zhang et al., 2006); privacy (Eichler, 2009; Engler, 2007);

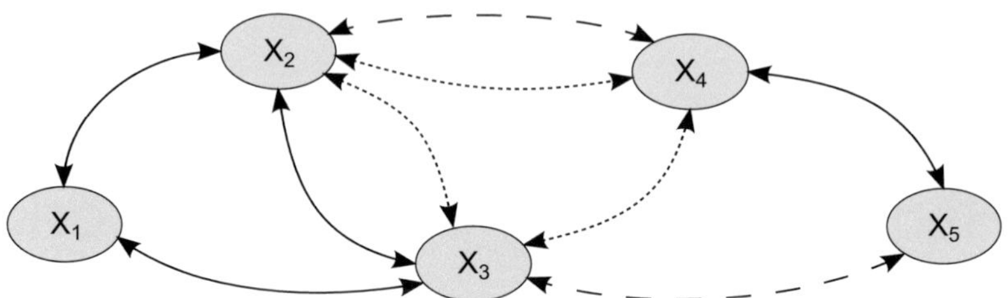

Figure 4.1: Common causes between several cooperation outcomes.

cryptographic mechanisms, like secure protocols or secure platform (Eichler, 2009; Engler, 2007; ISO/IEC11889-1, 2009; Zhang et al., 2006).

Reputation and recommendation systems work on the society-level as well. But they share some features with an individual-level trust mechanism. For this reason, the following references are used in this chapter too. The reputation system of the eBay Inc. (2013) is well known. It is a centralised system for a single platform. Another paper with a similar goal is Jøsang and Ismail, 2002. On the other hand, distributed reputation systems want to manage reputation across platforms and with enhanced privacy (Engler, 2007; Yu and Singh, 2002).

4.1 The Causality of Trust Development

The trust mechanism forms an expectation about an act of cooperation. Such an expectation is a glance in the future, a prediction of how the interaction partner will behave. When removing all the logical constraints of what can happen, a trust situation remains uncertain though, because the cooperation partner is there with his freedom to act. There is a lack of control or information as already discussed in Section 2.1. Why is it possible then to take a cautious look into the future and predict the other's freely chosen behaviour? Prediction is only possible, if the other one acts not completely random. There must be statistical dependencies in the behaviour of the cooperation partner. This is the reason, why developing trust makes sense at all.

For example, if vehicles cooperate in a network by exchanging information, a cooperation outcome could be represented as the difference between the reported and the real value. When a vehicle received five messages in the past, the cooperation outcomes can be denoted by the random variables $X_1, \dots, X_5$ as shown in Figure 4.1. Some of these outcomes could be statistically dependent. The messages of the outcomes X_1, X_2 and X_3, for example, could have been from the same sender, and those of the outcomes X_4 and X_5 from another sender. A sender has certain sensors and software with their own abilities and drawbacks. They could induce sender-specific correlations between the outcomes. In the figure, the double arrows with a solid line indicate that the messages of the connected outcome variables have been sent by the same trustee. Similarly two vehicles could have observed the same traffic sign; they performed the same task. Some tasks are more difficult or require different abilities than others. So the task as the subject of cooperation could induce a statistical dependency too. In the figure, this dependency is represented by a dashed line. For example, the messages two and four, and the messages three and five could refer to the same task. Finally the situational context, in which the message is received, could correlate with the cooperation outcome as well. Some regions, for example could have lower criminal activities or, simply, better cars than others. In the figure, the messages two, three and four have been received in the same situational context. The correlation between their outcomes is shown with dotted arrows. So a double arrow in the graphic indicates that the connected variables are correlated with each other but do not influence one another. Instead a common cause that is hidden affects both. (Pearl, 2000 introduces such graphical notations of causality.)

In the example, the sender can be characterised by some attributes like, whether it is an emergency vehicle, how old it is, or whether it is a bus, truck, etc. Most of these attributes do not directly affect the cooperation outcome, but they may be correlated with it. The real cause may be hidden in the sensors and the software of the sender. The sensory system, in turn, comes from the manufacturer, which also influences all other attributes. So there are hidden causes that connect the attributes of the sender with cooperation outcomes and the outcomes with each other. The resulting cause-effect chain goes from the hidden variables to the cooperation outcomes and other observable attributes, but not from the attributes to the cooperation outcome.

Note that some observable variables may influence other observable variables. There is a direct cause-effect relation. These cases are more obvious and easier to understand than the above consideration though.

In summary, the hidden variables introduce statistical dependencies between the observable variables. The dependencies in turn make it possible to predict some observable variables from others. *Trust is a mechanism that exploits these statistical dependencies.* With trust, a system can infer from some observable variables to the cooperation outcome based on past interactions. But what are these observable inputs?

4.2 Influences

If a trust mechanism should handle the uncertainty of a trust situation, it must assess this situation comprehensively to understand it. This section shows what attributes of a trust situation influence the trust development in the literature. I do not state every single attribute, which the various trust methods consider, but group them by entities they belong to, like the cooperation partner or the cooperation task.

Some basic trust algorithms attribute trust completely to the cooperation partner (Bamberger et al., 2010; eBay Inc., 2013; Zhang et al., 2006; Jøsang and Ismail, 2002; Yu and Singh, 2002). They disregard that a cooperation partner could perform various tasks differently. The algorithms take all outcomes from past interactions with a certain partner, rate them and estimate the trust in a new act of cooperation based on these ratings.

If a method develops trust depending on the task, it is often called context-aware. Such methods usually derive trust from the outcome ratings too, but incorporate some degree of interdependence between tasks (Bamberger et al., 2012; Engler, 2007; Ramchurn, 2004; Sabater and Sierra, 2002). The tasks are pre-defined as a list. The methods do not consider task and partner attributes. There are two exceptions though: Firstly, Ramchurn (2004) discusses the role of norms and rules (e.g. legal or platform dependent), which are attributes of the task. Secondly, Sabater and Sierra (2002) incorporate an attribute of the cooperation partner to model that the trustee may belong to a group or institution.

Golle et al. (2004) as well as Raya et al. (2008) take a completely different approach. They disregard or neglect the possibility of learning from past cooperation outcomes, mainly because they think that the interaction partners do not cooperate several times. Instead they only consider the plausibility of the exchanged information. The algorithms reason about the geometric setting, possible sensor capabilities, the type of information or a group-membership of the partner, for example. So they relate present attributes of the task and the partner entity with each other.

Only the method of Rettinger et al. (2008) handles the available information in a generic way. It explicitly mentions the entities for the cooperation partner and the state of the context. It considers a set of characterising attributes for both of them and sets of attributes that describe the cooperation agreement and the real cooperation outcome. However their method does not consider attributes of the partner-task relation that are not part of the agreement or the real outcome. For example, Raya et al. (2008) use the temporal and geographical proximity between the task and the cooperation partner to deduce the plausibility of the received data.

So far, the first group of methods learns trust from past cooperation outcomes. The second group of methods considers attributes of the present situation only to derive something that could be called the plausibility of some information. The work of Castelfranchi and Falcone forms a third line of theory regarding the influences of trust. (The following introduction specifically refers to the method described in Falcone et al., 2003.) They propose a high-level method that works on subjective beliefs, not on observable attributes of the trust situation. Their nomenclature is strongly influenced by work in the social sciences.

Following their approach, the following five beliefs mainly determine the trust in a given situation. The first three are attributed to the cooperation partner (internal factors); the last two regard the cooperation task (external factors).

The ability/competence belief describes the partner's physical and mental skills to choose and perform the right procedure in order to reach the desired goal. For example, to reload some goods, a robot needs specific arms, a sufficient allowed maximum load, and the algorithms to control his hardware sophisticatedly.

The willingness/disposition belief refers to the willingness, steadiness and availability of the trustee to act as agreed. The ability for an action is insufficient. The partner must also come into action. Certainly the trustee's behaviour may be an indicator for his willingness. But also the fact that he belongs to a certain group could indicate his willingness and steadiness, especially if experiences are missing. Falcone et al. attribute willingness to cognitive agents only. I take the term for all kinds of technical systems in this dissertation.

The unharmfulness belief represents the absence of risk that is intrinsic to the cooperation partner. This risk includes, for example, the danger of a machine break. So the unharmfulness belief addresses the quality of the work like the ability belief, but it emphasis the negative experiences. It is the belief that relates to the feeling of safety with regard to the cooperation partner.

The opportunity belief summarises opportunities of the context that are independent of the cooperation partner. If a vehicle, for example, reports a traffic sign, the fact whether the sign has been observed at night (little light) or during the day (much light) may be relevant context information.

The danger belief covers the absence of obstacles that are independent of the cooperation partner. For example, when a medical assistant during his education treats you, you may feel safer if an experienced doctor is nearby. The basic task and person is the same, but the risk assessment changes because of this additional context information.

These beliefs together provoke trust. Falcone et al. propose a scheme based on a fuzzy cognitive map to combine the beliefs to a trust belief. But were do those beliefs come from? How can a technical cognitive system judge the willingness of the cooperation partner? Falcone et al. leave this question open; they just name four general kinds of sources:

Direct experiences cover former trust situations whose outcome is already known.

Categorisation of the cooperation partner refers to the deduction of class properties to a class member. For example, a doctor could be trusted without knowing him directly, just because he is a doctor.

Reasoning is a more general form of deduction than categorisation. And finally,

Reputation covers the experiences of others.

In their evaluations, these sources have predefined values. All in all, Castelfranchi and Falcone seem to model human trust and apply it to multi-agent systems, specifically to belief-desire-intention agents (Wooldridge, 2000). While the author of this dissertation thinks that there are some main differences between inter-personal trust and inter-machine trust (as discussed in Chapters 2 and 3), the work of Castelfranchi and Falcone gives an important alternative view on trust in distinction to the other literature introduced in this chapter.

This section covered three approaches to trust theory: One focuses on the rating of past cooperation outcomes. Another one checks the plausibility of the present situation. And the third line evaluates beliefs on a high abstraction level. In the following, the various considerations in the literature are systematised and extended with own ideas. They all regard the same problem but from different points of view.

View on the affecting entities. Many trust algorithms rate cooperation outcomes depending on the cooperation partner and the performed task. In contrast, the vehicular network example, which was described in the previous section, shows that there are two kinds of situations: that of the task execution and that of the message exchange (the bargaining). Thus there are indeed three entities affecting the trust considerations: The cooperation partner, the task context and the bargaining context. They are further detailed in Section 6.2.

View on correlating variables. The previous section already showed that the cooperation outcome correlates with some other attributes of the trust situation. Consequently all attributes of the mentioned entities that may correlate with the cooperation outcome should be included in the inference to exploit all available correlations.

Temporal view. All entities induce dependencies in the outcomes of several interactions. This makes it possible to learn from experiences. The present situation also restricts the possible states of the future. Consequently the constraints in the presence allow to judge the plausibility of possible cooperation outcomes. (How the plausibility is checked, could be learned from experience in turn.)

View on the partner's abilities. This view matches the partner with the trust situation from the trustor's point of view. The trust situation exposes a problem to the trusting system and the other system is the right partner, if it is able to help with the problem in the desired way. So the ability refers to the view of the trustor on the trust situation. (Here the term ability refers to expertise but also to traits like empathy, loyalty, etc. They can all help to solve a personal or professional problem.) As a consequence, a trust algorithm should be based on all observable variables that could correlate with the partner's abilities.

View on the partner's willingness. This view matches the partner's needs and chances with the trust situation as perceived by the trustor. The trustee faces the trust situation and relates it to its own goals. So the willingness refers to the view of the trustee on the trust situation. It affects the cooperation outcome as well. As a consequence, a trust algorithm should regard all observable variables that could correlate with the partner's willingness. Especially during the bargaining, the partner could express its willingness.

All these views help to understand the inference problem completely. The following requirement summarises that.

Requirement 1 (Influences). To judge an act of cooperation, a trust method should evaluate it comprehensively. Thinking

- of the affected entities, their attributes and correlations;

- of the different situations the trusting system has experienced over time; and

- of the abilities and willingness of the trustee

helps the algorithm designer to identify all influences. In addition, the expectation computed with a trust algorithm should indeed depend on those influences.

As a main contribution, this document unifies all these lines of trust theory in one model of trust, the Enfident Model. This model can be understood as a framework to analyse and understand existing trust algorithms as well as to devise a new trust algorithm specific for a certain application.

4.3 Output of a Trust Algorithm

This section describes some qualitative properties of the output that should be provided by a trust algorithm. The right output depends on how it is processed in subsequent modules.

4.3.1 Trust Representation

All investigated trust algorithms provide a trust value. It may be a discrete or continuous value (often in the ranges [0, 1] or [−1, 1]); and it is computed for every combination of input parameters. So some algorithms distinguish only one trust value per cooperation partner, others attribute many trust values to a cooperation partner, depending on the input parameters.

Ramchurn (2004) as well as Sabater and Sierra (2002) emphasise that every single property of a cooperation outcome should be evaluated separately instead of one rating per outcome. In the scenario of a virtual market place for example, the delivery time, the properties of the delivered product and the price in the bill should all be rated on their own. This way, a decision algorithm can better balance the ratings depending on the current situation.

Some trust algorithms provide just a trust value (Falcone et al., 2003; Golle et al., 2004; Jøsang and Ismail, 2002; Ramchurn, 2004; Raya et al., 2008; Rettinger et al., 2008). Others associate a trust value with an additional ignorance value (Bamberger et al., 2010, 2012; Engler, 2007; Zhang et al., 2006). The ignorance value describes how certain the trust value is. So it is usually related to the amount of evidence, on which the trust value is based.

For some authors, a trust value fails to sufficiently describe the past behaviour of the cooperation partner. As a consequence, Yu and Singh (2002) introduce an additional distrust value to better distinguish positive experiences (the trust value) and negative experiences (the distrust value). Sabater and Sierra (2002) incorporate a deviation value that should reflect the variability (or unsteadiness) of the partner's rating.

Both, the distrust value and the deviation value, try to better describe the form of the distribution over the cooperation outcomes. But again, they fail to represent any kind of distribution. They still reduce the information contained in a distribution into two values. So in general, a decision algorithm could benefit from knowing the whole distribution instead of just these two values.

For this reason, I propose that a trust algorithm should provide the whole distribution over the cooperation outcomes, the *trust distribution*; and it should do so for every single property of the outcome (see also Section 2.5).

A distribution is a general description, from which various kinds of trust values can be derived. It is also very useful for a decision algorithm, because it can be well combined with a utility function and other decision-related considerations.

In addition, a trust distribution should be accompanied by an ignorance value to quantify the evidence the trust distribution is based on. Both together provide the complete information from which other characterising measures can be derived.

Requirement 2 (Trust representation). A trust algorithm should compute an expectation for every attribute of a cooperation outcome separately. An expectation should quantify the belief in or the likelihood of every possible value of its attribute (the trust distribution). In addition, the algorithm should state the certainty of the trust distribution, for example, in form of variances for the distribution parameters or in form of the amount of evidence.

A trust algorithm can also provide a trust value if needed and a distrust value. Both typically characterise some aspects of the trust distribution and can be derived from it. They are especially useful to talk about trust with other systems.

In the same way as the trusting system should evaluate the likelihood of every possible cooperation outcome, the trusted system should be allowed to give likelihoods for the possible cooperation outcomes during the bargaining. A vehicle, for example, could say that there is a speed limit sign showing the limit 80 km/h with a certainty of 0.2 and 60 km/h with a certainty of 0.8. Prohibiting uncertain outcome agreements would suppress the spreading of uncertain information although such information can be useful for the receiver.

4.3.2 Trust and Decision

Some trust methods include the decision logic (Raya et al., 2008; Zhang et al., 2006); thus they integrate various pieces of information to one value or determine the optimal action to perform. In contrast, the Enfident Model and many other trust methods stop directly before the decision making.

In alignment with the literature in the social sciences (Sections 2.5 and 3.1), I think that decision making is influenced by much more than trust. In other words, trust is an attitude that can manifest itself in corresponding behaviour but needs not. So I expect decision algorithms that exploit the information provided by a trust algorithm, but also consider further aspects of the situation like other sub-goals or alternative sets of actions. As a consequence, the decision algorithm should be separate from the trust algorithm. (Compare this with the systematics in Section 2.5.)

Requirement 3 (Inner state). A trust algorithm should provide an expectation towards a certain cooperation outcome. This expectation represents an inner state of the system. It needs not determine its final action.

So from a conceptual perspective, trust development and deciding are two distinct concepts. But of course, it could be reasonable to combine both in one algorithm. This depends on the application.

4.3.3 Trust, Risk and Utility

The provided trust value can be understood as a likelihood indicator (as in Bamberger et al., 2012, 2010; Raya et al., 2008; Rettinger et al., 2008; Zhang et al., 2006; Golle et al., 2004) or as a utility (or risk) indicator (as in Ramchurn, 2004; Sabater and Sierra, 2002). (Certainly in some scenarios, both are equivalent.) This distinction of a likelihood and utility/risk interpretation applies only to trust methods that have a technical system as the evaluating entity. For reputation methods with human user input, the interpretation as a likelihood or utility is up to the user (Engler, 2007; Falcone et al., 2003; Jøsang and Ismail, 2002; Yu and Singh, 2002).

I think, whether the trust value should reflect the utility and risk (interest-related trust) or just a rate (probabilistic trust), depends on its usage. (Compare this with the systematics of Section 2.5.) For this reason, the Enfident Model addresses both cases. Considerations about utility and risk are typically located in a decision module, which is much more than evaluating trust. Therefore the probabilistic trust distribution is recommended for the interface between the trust module and the decision module. On the other hand, an interest-related trust value could be useful for talking about trust and interacting with reputation systems.

4.4 Reasoning Process

So far, the input and the output variables have been identified. They already say much about what a trust algorithm should do. In the following, some general aspects are discussed how a trust algorithm should reason on the available evidence. They are qualitative aspects independent of any specific algorithmic technique.

4.4.1 Present Constraints and Experience

Golle et al. (2004) as well as Raya et al. (2008) consider only parameters of the current situation and combine them with a fixed algorithm that discards experiences. (Refer especially to p. 1916 of Raya et al., 2008.) Besides other techniques, they exploit logical constraints between the variables to judge the cooperation outcome (see also the previous Section 4.2).

Falcone et al. (2003) combine beliefs that may origin in experiences. They define a fixed logic for the combination. They never state though, how to obtain these beliefs.

Most methods explicitly show how to use past outcome ratings to estimate the trust in new cooperation (Bamberger et al., 2010, 2012; Engler, 2007; Ramchurn, 2004; Jøsang and Ismail, 2002; Sabater and Sierra, 2002; Yu and Singh, 2002). The way how they combine these ratings is a logic with fixed parameters though. Such an algorithm requires the designer to know exact correlations and thresholds. This is often not the case.

Zhang et al. (2006) make a first step towards parameter learning. They use clustering to estimate one parameter. Only Rettinger et al. (2008) propose a complete statistical model, for which most parameters can be learned. He uses a technique similar to that proposed in Chapter 7.

All in all, I think that learning from the past is inherent to the trust mechanism. Experiences can help to better estimate trust in analogue situations. Together with generalisation techniques (see below), they also help to judge about new situations. Additional logic also reduces the number and the probability of future states. Thus present constraints are another important tool to compute an expectation about the cooperation outcome.

How certain experiences and certain attributes of the current situation cause a new cooperation outcome, is hard to define. It may even depend on the specific environment the system lives in. As a consequence, a trust method should define a schema for learning (of parameters and, possibly, even of the reasoning structure).

Requirement 4 (Present constraints and experience). The reasoning process of a trust algorithm should evaluate constraints of the current situation and learn from past acts of cooperation. The experience should even influence the parameters of the process.

The trustor should use the most up-to-date set of experiences and evaluate trust as late as possible. This is important because reasoning is non-monotonic, that is, additional evidence can change the reasoning result significantly (Pearl, 1988, p. 17). Use Case 6 on page 166 gives an example. Even the order of the experiences matters in some cases as Section 4.4.3 below discusses. On the other hand, all experiences cannot be stored forever. So efficient experience storage and late evaluation must be balanced reasonably.

Requirement 5 (Non-monotonic reasoning). A trust algorithm should be able to revise the judgement of a previous cooperation outcome.

4.4.2 Specialisation and Generalisation

A trust algorithm should learn from experiences to better judge a certain situation. It specialises on that situation with an increasing amount of experience. But a main purpose of trust is to support acting under uncertainty, that is, in quite new situations. Thus a trust algorithm should also generalise over all past situations to judge a new (possibly similar) situation. It should transfer the lessons learned so far to a new problem. So transfer learning (e.g., Pan and Yang, 2010) is an important technique for a trust algorithm to handle the uncertainty of new situations.

Those trust algorithms that do not learn cannot specialise and generalise (Raya et al., 2008; Golle et al., 2004; Falcone et al., 2003).

Some trust algorithms consider the cooperation partner the only situational property (Bamberger et al., 2010; Zhang et al., 2006; Jøsang and Ismail, 2002; Yu and Singh, 2002). So they specialise for the partner but for no further aspect of a situation. None of them considers generalisation.

Other algorithms allow different kinds of cooperation tasks (Bamberger et al., 2012; Engler, 2007; Ramchurn, 2004; Sabater and Sierra, 2002). So they specialise on the combination of the task class and the partner. But the classification of the tasks is predetermined, so the degree of specialisation is fixed. The algorithms handle generalisation in different ways. Sabater and Sierra (2002) as well as Ramchurn (2004) make no mention of generalisation. But in fact, different kinds of cooperation may share outcome attributes. So some attributes are learned across various trust situations. In contrast, Engler (2007) explicitly defines context area dependencies through a weighted graph. The weights must be known by the designer; so this form of generalisation is static. The algorithm of Bamberger et al. (2012) generalises across cooperation tasks dynamically. The experiences with other tasks should only fill the lack of knowledge about the task of interest. So if many experiences are available for a task, the experiences with other tasks only have a low weight. And the other way around, if only little is known about a task, the experiences with other tasks dominate the result.

Finally Rettinger et al. (2008) use probabilistic clustering for dynamic specialisation and generalisation (similarly to the algorithm proposed in Chapter 7). With an increasing number of experiences, the clusters have stronger characteristics (specialisation), while the number of clusters possibly increases. The clustering assigns every single experience to every cluster with a degree of similarity. So the algorithm generalises over all clusters. This is, to our knowledge, the only related work that can generalise over cooperation partners.

I think that handling uncertain situations is an important purpose of a trust algorithm. Generalisation helps to do so. It is a form of transfer learning. With it, the trusting system can judge cooperation partners, bargaining contexts or task contexts, even if no or only few experiences are available for some of them. So it especially supports to form initial trust as Rettinger et al. (2008) emphasise. (Initial trust is the trust in a situation, for which no experience is available; that means no experience is available for the exact combination of all three entities.)

Requirement 6 (Specialisation and generalisation). A trust algorithm should generalise over and specialise for the cooperation partner, the bargaining context and the task context.

Experiences that are more similar with a trust situation should stronger influence the resulting trust. This implies that the algorithm specialises stronger for a certain situation, the more similar experiences it knows.

For example, a system wants to judge a new trust situation that features the cooperation partner P_2 and the task context T_2. (For simplicity, I omit the bargaining context here.) It never had cooperation with the same combination of entities. But it cooperated with the partner P_1 for the task context T_2 and it cooperated with the partner P_2 for the task context T_1. Because P_1 and P_2 are similar to a certain degree as well as T_1 and T_2, the trusting system can generalise over the partner and the task entity at the same time to judge this new situation. In fact, it would consider all task contexts and all cooperation partners depending on their similarity to the current problem. Note that Zacharia and Maes (2000) require that a new cooperation partner should have an initial trust value that is related to, how easy an identity can be changed in the system. Otherwise a system with a bad reputation could simply change its identity to increase its reputation.

4.4.3 Entities as a Time-Varying Process

The cooperation partner can be understood as a statistical process, because its behaviour can be perceived as non-deterministic from the observer's point of view. So far, this chapter assumed that the inner working of the trustee does not change. The observable behaviour looks as it would come from a static statistical process.

While the basic set up of the cooperation partner may remain stable throughout its life time, the whole performance can change though. For example, a software update can improve the cooperation quality significantly. (It is more likely than a hardware modification). Another example for a non-static behaviour is a hardware defect. It results in a temporary decrease of the performance. After the defect has been repaired, the performance goes up again. This happens abruptly. A third notable example is malicious software that misuses the technical system for other purposes. This form of unwanted system modification is especially discussed in the field of cognitive vehicles that cooperate through a vehicular network. Section 8.2 gives exemplary use cases for it.

So assuming the process to be static is inappropriate. The model must include a *concept of time* that allows the process to affect the cooperation outcome differently depending on when the cooperation happens.

The examples above showed that the cooperation partner must obviously be modelled as a time-varying process. Is this also the case for the other entities of Section 4.2 (the bargaining context and the task context)? On a first glance, a situational context describes something that has no own way of working. It is just a collection of variables, which the participating systems perceive. But this is exactly, why a situational context may change its influence on the cooperation outcome, although its observable attributes do not change: Either the trustor or the trustee perceives or evaluates a situation differently than before, or some situational properties that are missing in the trust algorithm have changed. So yes, basically all entities can be time-varying processes. However this way of modelling seems especially important for the cooperation partner.

Requirement 7 (Time-Varying Processes). A trust algorithm should respect time and treat the entities of a trust situation like time-varying processes. In the application, this could be unnecessary for some entities though.

How is this requirement realised in the literature? Some algorithms disregard experiences and thus do not model the dynamics of the entities (Raya et al., 2008; Golle et al., 2004; Falcone et al., 2003). Other algorithms take the experiences as an unordered collection (Rettinger et al., 2008; Zhang et al., 2006; Yu and Singh, 2002). Consequently they neglect time as well. While Ramchurn also disregards the temporal order of the experiences, he discusses that the algorithm could better adapt to the current behaviour of the cooperation partner if it drops old experiences (Ramchurn, 2004, pp. 76, 101 and 129)

Some algorithms take a step further by weighting the experiences according to their ages (Engler, 2007; Jøsang and Ismail, 2002; Sabater and Sierra, 2002). This is a simple but explicit model of the time-dependent nature of trust development. Bamberger et al. (2010, 2012) address quick changes in the behaviour of the cooperation partner with loop control. The mechanism is too simple for complex scenarios though. Only recently (January 2013), I found a paper (ElSalamouny et al., 2010) with a similar approach like mine for this requirement. There a hidden Markov model (see Bishop, 2007 for an introduction) produces the outcome of an interaction. But the model is still to simple to integrate other trust functionalities like those of Requirements 1, 4 and 6.

4.5 Summary

This chapter looked at various aspects of a trust algorithm, from the input variables and the processing of the algorithm to the output variables. So it followed again the schema of an input-output system. The focus was on the trust development in the individual. Other trust-supporting features were omitted. Note that the input-output schema is not used for the Enfident Model in the next Chapter, because it is too restricted for a general reasoning approach as Section 6.4 points out.

The work of other authors and own ideas complemented each other to get an image of, what a trust method should do, independently of a specific application. Seven requirements have been identified. Some of them are not met by any current trust algorithm. Such a definition of clear requirements for trust methods is new. It is a contribution to a theory of trust between technical systems.

5 Notation

This chapter clarifies the notation and some terms, which are used throughout this dissertation. This is done on an example. As a consequence, the
text sometimes reads like a tutorial, although it does not introduce the subject matter. The text just introduces the notation. For further readings on the
subject matters, references are given in every section.

5.1 Probabilistic Notation

This section describes the statistics-related symbols and notation that are
used in the remaining chapters. The reader can find an introduction to
modern probability theory based on measure theory, for example, in Bartoszyński and Niewiadomska-Bugaj, 2008; Fristedt and Gray, 1997 and
Schmidt, 2009. This section is based on these books.

I introduce the notation on an example. This makes it easy to understand
the symbols, even if the reader knows a different notation. In the exemplary
experiment, a coin is tossed twice. It can show its head (referred by the
symbol H) and its tail (referred by the symbol T).

Probability space. The *sample space* $\Omega = \{HH, HT, TH, TT\}$ is the set
of all outcomes. Any subset of the sample space is called an *event*. For
example, $a = \{HH, HT\}$ is the event that the first toss shows the head.
The event corresponding to the empty set is called the *impossible event* or
null event. The *probability measure P* assigns a probability to every event
including the null event ($P : \mathcal{F} \mapsto [0, 1]$). $\mathcal{F}$ is a σ-field or σ-algebra of
subsets of Ω. (It "is a collection [...] of subsets of Ω that has $\emptyset$ as a member
and is closed under complementation and countable unions" (Fristedt and

$f \in \mathcal{F}$	$P(f)$	$f \in \mathcal{F}$	$P(f)$	$f \in \mathcal{F}$	$P(f)$	$f \in \mathcal{F}$	$P(f)$
$\emptyset$	0.00						
HH	0.25	HT	0.25	TH	0.25	TT	0.25
HH, HT	0.50	HH, TH	0.50	HH, TT	0.50		
HT, TH	0.50	HT, TT	0.50	TH, TT	0.50		
HH, HT, TH	0.75	HH, HT, TT	0.75	HH, TH, TT	0.75	HT, TH, TT	0.75
HH, HT, TH, TT	1.00						

Table 5.1: The probability table that describes the probability measure P in the coin example. An event f has happened if the experiment's outcome is part of it.

Gray, 1997, p. 6).) The pair $(\Omega, \mathcal{F})$ of a set Ω and a σ-field $\mathcal{F}$ over Ω is called a measurable space. The triple $(\Omega, \mathcal{F}, P)$ forms the *probability space*. Table 5.1 describes $\mathcal{F}$ and P for the coin example.

An intersection of events appears often in the text. To simplify the notation, I write commas for this operation. This is very common in the literature. So the joint probability $P(a, b)$ is equivalent to $P(a \cap b)$.

Random variable. A *random variable* X is a mean to describe an event. Its value can be considered the result from a measurement of a random process. Technically it is a measurable function from the probability space $(\Omega, \mathcal{F}, P)$ to another measurable space $(\Psi, \mathcal{G})$, so $X : \mathcal{F} \mapsto \mathcal{G}$. (Note that some authors further restrict random variables to have numerical values. I do not so.) For example with the help of a real-valued random variable X, the inequality $X < a$ describes an event in the image of X. In the domain of X, the same event is represented by the set of all elements $\omega \in \Omega$ that satisfy $X(\omega) < a$. In the coin example, the random variable X could assign a number between 1 and 4 to every element in Ω ($\Psi = \{1, 2, 3, 4\}$). Then $X = 3$ corresponds to the event $\{TH\}$ with the probability $P(X^{-1}(\{3\})) = 0.25$ and $2 \leq X \leq 4$ corresponds to the event $\{HT, TH, TT\}$ with the probability $P(X^{-1}(\{2, 3, 4\})) = 0.75$. ($X^{-1}$ is the preimage of X.) So random variables can be used to easily name certain events. As another example of notation, the random variable Y could give the result of the first coin toss ($\Psi = \{H, T\}$). Then $P(Y = H)$ refers to the probability of the event $\{HH, HT\}$.

To ease writing the probabilities, another probability measure $P_X : \Psi \mapsto [0, 1]$ could be introduced and defined as $P_X(\psi) = P(X^{-1}(\psi))$. It is the *distribution* of X. Together with P_X, the image space $(\Psi, \mathcal{G})$ is a probability space as well. The notation $X \sim F$ expresses that the random variable X has the distribution F.

Sometimes this text uses the term *discrete random variable*. It refers to a random variable from Ω to Ψ, for which a countable set $U \subseteq \Psi$ exists with $P(U) = 1$.

In addition to the strict notation above, this dissertation uses the following common simplifications. The symbol P denotes arbitrary probability measures, although they are different objects. The context and the argument of the measure help to distinguish them. Furthermore events $\{\omega\}$ containing only a single outcome $\omega \in \Omega$ can be written without surrounding brackets. Consequently if two experiments are considered, the double coin toss of the example above and a single coin toss associated with the random variable Y, then the following notation can be used: $P(\{HH\}) = P(HH) = P(X = 1) = P(\{1\}) = P(1)$, $P(\{Y = 1\} \cap \{X = 2\}) = P(Y = 1, X = 2)$ and "$P(Y = 1)$ is greater than $P(X = 1)$".

All in all, random variables are widely used to give names to events. They are usually defined by their image, hiding the underlying probability space. This approach aligns well with the invariance principle of random variables postulated by Bartoszyński and Niewiadomska-Bugaj (2008, p. 126). They state that a single experiment can usually be described with several sample spaces. Random variables help to work independently of the underlying sample space. *In this dissertation, the sample space often remains undefined. Instead the image of a random variable is regarded.*

Conditional probability and independence.   When, in the running example, the coin is tossed once and it shows its head, it is said that the event $c = \{HH, HT\}$ occurred. Knowing this event changes what probability is associated with the events of the probability space. The old probability measure becomes invalid, because some additional information is available. The experiment changed. The new probability measure is called the *conditional probability measure*. It is written as $P(a \mid c)$, where a is a free variable defined on $\mathcal{F}$ and c is fixed. Schmidt (2009) defines the conditional probability measure as the indefinite Lebesgue integral

$$\int \frac{\chi_c}{P(c)} \, dP, \qquad \text{provided } P(c) > 0.$$

$\chi_c(\omega)$ is the indicator function of the event c. It is 1, if $\omega \in c$, and 0 otherwise. The value of the conditional probability measure for a certain event a is called the *conditional probability of a given c*. It can be computed with the well known formula for the conditional probability:

$$P(a \mid c) = \int_a \frac{\chi_c}{P(c)} \, dP = \frac{P(a \cap c)}{P(c)}. \tag{5.1}$$

Inference is based on partial knowledge (experience) about the future state of the world. This partial knowledge can be modelled as a condition in the way shown above. So the idea of conditioning will be very important throughout this dissertation.

In modelling tasks, it is sometimes hard to quantify probability measures. In contrast, the following independence properties can often be found intuitively. For this reason, I recall them here; they are important for the following sections and chapters.

Two events a and b are *statistically independent*, if the probability of the one does not change in the light of the information that the other one has occurred. Formally this is the case, if the equation

$$P(a \cap b) = P(a) \, P(b)$$

holds. Often the wording is abbreviated by saying a and b are independent – omitting the word "statistically".

Let h be another event with $P(h) > 0$. Then a and b are *conditionally independent given the event h*, if

$$P(a \cap b \mid h) = P(a \mid h) \, P(b \mid h).$$

This means, when knowing h, the information about B does not change the probability of a. Finally note that both independence properties above are a commutative property of the relation. They hold in both directions: a is independent of b and b is independent of a.

Knowing that two events are statistically or conditionally independent usually simplifies a model. This is the reason, why both properties are so important for this dissertation.

Finally I recall the well known theorem of Bayes, here in the form that is often considered in statistical inference. The theorem follows from Equation 5.1.

$$P(A = v_i \mid d, \mathcal{H}) = \frac{P(d \mid A = v_i, \mathcal{H})\, P(A = v_i \mid \mathcal{H})}{\sum_{k=1}^{N_A} \left[P(d \mid A = v_k, \mathcal{H})\, P(A = v_k \mid \mathcal{H}) \right]}$$
$$\propto P(d \mid A = v_i, \mathcal{H})\, P(A = v_i \mid \mathcal{H})$$

The symbol A denotes a random variable, which can take on the values v_i with $i = 1, \dots, N_A$. d represents the observed data and $\mathcal{H}$ the hypothesis, which typically contains the parameters and the model structure. The symbol $\propto$ expresses that the right hand side equals the left hand side except for a scaling factor that is independent of v_i. In other words, normalising the right hand side so that it sums up to one results in the left hand side. I call $P(A = v_i \mid \mathcal{H})$ the *prior distribution* or just the prior, $P(A = v_i \mid d, \mathcal{H})$ the *posterior distribution* or just the posterior and $P(d \mid A = v_i, \mathcal{H})$ the *likelihood*, which *updates* the prior in the light of the data.

5.2 Graphical Notation

To describe the problem of trust development, I need a notation that combines data with uncertainty and logic. This dissertation uses the probabilistic entity-relationship models of Heckerman et al. (2007). The directed form of this graphical language combines elements from entity-relationship models (Chen, 1976) and Bayesian networks (Pearl, 1988). For this reason, it should be easily understandable for readers that know those. Here I only recall the elements that are used in the dissertation to clarify the notation. For an introduction, please refer to Heckerman et al., 2007.

Entities refer to subjects and objects of the model and help to identify the data of interest. Relationships describe the logic that underlies the entities. Entity-relationship models focus on the types of entities, not on entities directly. As an example, consider a bag of coins, every one of which is tossed twice. Then each experiment consists of three entities: one coin and two tosses in a specific order. With every coin, the same two tosses are performed. They are the single trials of the experiment. So every coin is in a relationship with two trials.

The exemplary setting can be transformed in a probabilistic entity-relationship model. It contains two entity types: the type *Coin* and the type *Trial*. Every entity of the type Coin is related to two entities of the type Trial. This relationship is of the type *Thrown*. Figure 5.1 shows how this setting can be expressed graphically as an entity-relationship model. An entity type is drawn as a rectangle and a relationship type as a diamond. They are connected with dashed lines. Observe that this model describes a logic on a data set. The entity types induce the predicates *is-coin* and *is-trial*. They help to test, whether an entity e is of a certain type: is-coin(e) is true only, if e is an entity of the type Coin. Similarly the relationship introduces a predicate *is-thrown-in* that connects coins and trials. If a trial t has been performed with a coin c, the predicate is-thrown-in(c, t) is true, otherwise it is false. (In this example, of course, that predicate is always true, because both trials have been performed with all coins.)

Figure 5.1: Graphical description of the entities and their relations in the coin toss example. The rectangles are entities; the diamond is a relationship.

So far, the model simply described the setting. Typically people are also interested in the data of an experiment. Imagine that a coin is white, blue or black and it can show head or tail. This introduces two attribute types.

Every entity of the type Coin has an attribute of type *Coin.Colour* and every relationship of the type Thrown has an attribute of type *Thrown.Side.* Through the entities and their relationships, every side value is associated with a colour value. Attribute types are drawn as ovals and connected with their entity or relationship types by a dashed line as shown in Figure 5.2. This diagram also contains the attribute type *Trial.Number* to distinguish the first from the second toss. Note that an attribute type can be understood as a function from a set of entities or relationships of a certain type to a value set.

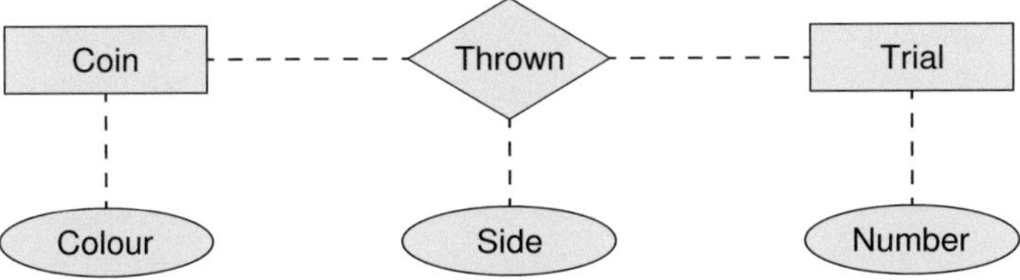

Figure 5.2: Graphical model of the entities, relationships and attributes in the coin toss example. Attributes are drawn as ovals.

A main point of an entity-relationship model should be emphasised here. The model describes the structure of data sets. To make a specific data set out of it, the entities and their relationships (the skeleton) as well as the values of the corresponding attributes must be given. Such a specific data set is an instance or a realisation of the model.

Entity and relationship types make up a type hierarchy as known from object-oriented programming languages. This is an important feature to model different but possibly similar trust situations. A type Y derived from X is associated with the same attribute and relationship types as X, but can have additional attribute and relationship types. Thus a sub-type extends the parent type. This way, the concept of a cooperation partner can be modelled as a type, while the sub-types are kinds of machines, which can have completely different attribute types. For example, a vacuum cleaning robot can be characterised by its year of manufacturing and the power of its engine, and a kitchen robot by its year of manufacturing and its kind of arm. Both types can exist in the same model and be handled as cooperation partners. I regard a type that is not derived from any other type as a *root type*.

The following three functions help to access the structure of the data. The function *entities* regards the entities associated with an attribute. If *a* is an attribute of the entity *e*, entities(*a*) maps to the set $\{e\}$. And if *a* is an attribute of the relationship *r*, entities(*a*) results in the set of all entities that are connected with *r*. The function *type* maps to the type of an entity, relationship or attribute *x*. It provides the specific type of an object, not a parent type. If *x* is of the type *t*, type(*x*) results in *t*. Finally the function *roottype* maps to the root type of the entity *e* in the type hierarchy. If *e* is of the type *t* which is derived from the root type *R*, then roottype(*e*) results in *R*. Note that all these functions take instances as arguments, not types.

Imagine now that the coins are not fair. Instead they have different binomial distributions depending on their colour. So the side attribute correlates with the colour attribute. This can be expressed with an arc type. This again is an element on the type level. It can be realised for given attributes. An arc type is drawn as an arrow in the diagram. The direction of the arrow specifies a conditional probability distribution. The starting point is the condition. If the conditioning does not matter, just a solid line with no arrow is used. *Note that the main element of this graphical notation is the absence of an arc. Attributes that are not connected are conditionally independent.* Figure 5.3 shows the arc type between the attribute types Coin.Colour and the Thrown.Side. Because the trials with a certain coin are statistically independent, no arrow starts at Trial.Number.

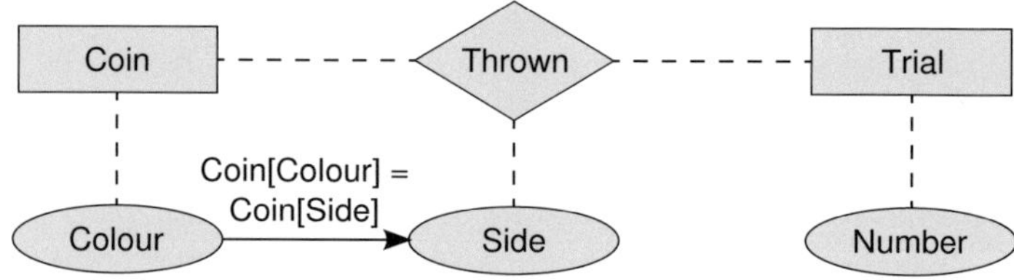

Figure 5.3: Probabilistic entity-relationship model of the coin toss example.

In the diagram, the arrow is annotated with "Coin[Colour] = Coin[Side]". This annotation is a constraint on the arc type. In a realisation of the model, the arc is only present, if the constraint is fulfilled. The given constraint in the figure is a sloppy form of the expression that for a given coin c and a given relationship Throw(c', t) between a coin c' and a trial t, the arc is present only, if $c = c'$. So this constraint simply says that an arc is not drawn between all colour and side values, but only between the coin's colour and the resulting side that were part of a single trial. Because it is the typical case in this document that arcs connect only the attributes of those entities and relationships that are either connected directly or through a relationship, that constraint is taken as an implicit constraint and usually omitted. This convention makes the graphical presentation cleaner.

The probabilistic entity-relationship model again describes the structure of a data set and the independence relations in the structure. So it is no Bayesian network. Rather, entities and their relationships as well as values of their attributes are necessary to transform the model in a Bayesian network. In this dissertation, the model is often conceptual though. *It describes the problem but not the algorithm.* Thus it is not intended simply to transform it in a Bayesian network and run standard algorithms on them. Rather the presented model could be realised with different Bayesian networks and algorithms. Chapter 7 shows as an example of, how the conceptual model of Chapter 6 could be realised in a Bayesian network structure.

This dissertation is mostly interested in causes and their effects, not in probability distributions. Although the description of causality was not the intention behind probabilistic entity-relationship models, their notation suits well for this purpose (see also Pearl, 2000). In this context, an arrow describes a cause-effect relation; it is not associated with a conditional probability distribution. The starting point of the arrow is the cause, the end point the effect. Such a graphical model is not a generative model any more. It could be realised by generative and discriminative algorithms. This is the way, how I use the graphical notation throughout Chapter 6.

6 The Enfident Model

This chapter presents a trust mechanism for technical systems by showing different views on it. All these views describe the same model of trust. They are sub-models. I call the complete model the *Enfident Model*. It is a qualitative model of trust, because it only states how the reasoning should work, but omits a specific reasoning algorithm. Chapter 7 shows a straight-forward implementation then. A qualitative model helps to understand and investigate the principles of trust and to evaluate various reasoning techniques for trust development. As a consequence, the Enfident Model can be thought of as a meta trust model or a framework for specific implementations.

The next section introduces the basic setting of a trust situation. Section 6.2 then defines the data that is involved in a trust situation. The way how a reasoning mechanism should connect these pieces of data is shown in Section 6.3. The sections up to there describe a model that contains trust, but they do not state what the actual output of the model is. Section 6.4 then presents a flexible mechanism, how to get something out of the model. Finally Section 6.5 gives some advise for designers who make an implementation for a certain application scenario.

6.1 Trust-Related Situations

Trust involves two parties: The one who is trusting, that is, who expects certain behaviour of another one; and the one, whom is trusted and who acts. Both interact with each other in different situations, depending on the application scenario.

Some of these situations feature the behaviour and abilities of the trusted system. They are the *situations of the task execution*. For example, imagine

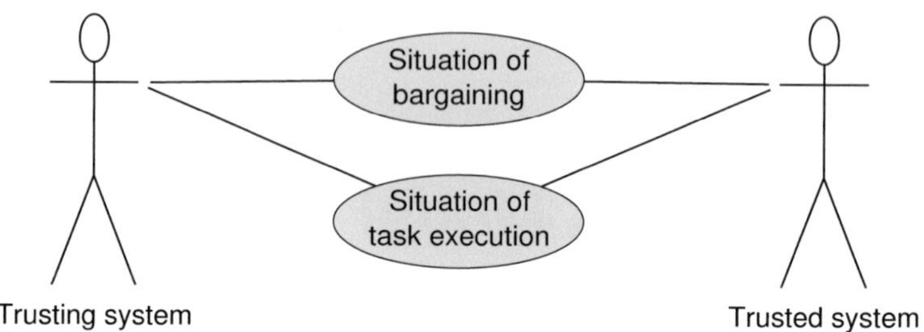

Figure 6.1: UML use case diagram of the basic trust setting. Two parties interact in two kinds of situations with one another.

the scenario of a logistics centre where three robots move a large box. One of them is the trusting system; the two others are trusted systems. They perform the task in cooperation. Or consider the scenario of information exchange in a vehicular network. The task here is to obtain some information that is exchanged later. So during the task execution, the trusted system may be at a different place than the trusting system and both may not even know each other yet. This setting may look unexpected at a first glance. It is detailed in Section 8.3. All in all, the situation of the task execution is characterised by the actions of the trusted system, about which the trusting system has an expectation. These actions require and, thus, manifest certain abilities of the trusted system.

In some scenarios, both parties negotiate on the intended act of cooperation before the task execution. This happens in the *situation of bargaining*. The following examples illustrate different forms of bargaining situations. In the logistics centre example, the robots may have talked about, what task is to do, whether the trusted robots can perform the task, when the task should be done, etc. In the vehicular network scenario, the bargaining happens after the task execution. The vehicles simply talk about information they have observed some time ago. In addition, the bargaining can be implicit here. The trusting vehicle often accepts any information with the only requirement that it is correct. Moreover consider a car that is overtaken by another one. The slower car is expecting and, thus, trusting that the faster car is not hitting it. This scenario of forced dependence between both vehicles has been introduced in Section 1.1. Here the cooperation is very loose and implicit. It

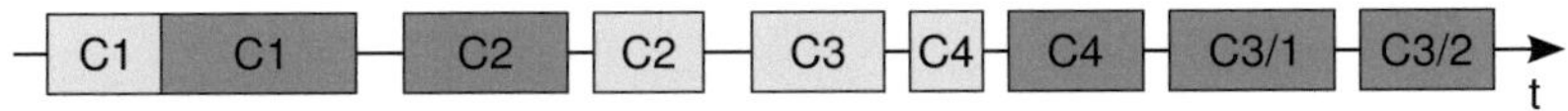

Figure 6.2: Exemplary time line of trust situations. C1, C2, C3 and C4 are acts of cooperation. The light grey refers to situations of bargaining, the dark grey to situations of task execution.

involves no explicit bargaining. In summary, while the bargaining does not feature the actions of interest, it may correlate with them. The subject of the negotiation connects the bargaining with the task execution. The trusted person can indicate its motivation and willingness in the situation of bargaining. This situation can be before or after the task execution. It can have the form of an explicit bargaining or it can be just an imagined situation of implicit bargaining.

As a result, the basic setting is that two parties, the trusting and the trusted person, interact in two trust-related phases. This simple but important idea is depicted in Figure 6.1 as a UML use case diagram. Both phases may consist of several single situations. For example, the performance of a task can take several days. Or one part of the cooperation could have been negotiated in one meeting, while other parts were added later. This view on the time is visualised in Figure 6.2. It shows that the bargaining can happen before the task execution (C1, C3 and C4) or after the task execution (C2). Every phase can be made up of several interactions, like the task execution of C3. And several acts of cooperation can overlap (C3 and C4).

When thinking about trust, the trusting system can be at any position on the time line: at the end of the bargaining, for example, or in the middle of the task execution. There it may reason about an unknown property that is obtained anywhere else on the time line. For example, it could reason about how moving some containers will end up – in the future –, or it could want to know what the sending vehicle has really observed – in the past – in contrast to what it has sent. This is the basic setting underlying the Enfident Model: two parties interact over time in situations of two kinds, accompanied by uncertainty. The model takes the available information and reasons about missing pieces. The next section systematises the involved information from an entity-relationship perspective.

6.2 Data Definition

This section defines the data, which appears in the Enfident Model, in an object-oriented way. A trust algorithm learns models of the others' behaviour. As a consequence, it must consider all information that can correlate with the behaviour of interest.

6.2.1 Overview

Figure 6.3 visualises the data model in form of an entity-relationship diagram (Chen, 1976). It describes types of objects and a logic that relates them with each other. The model is detailed in the following.

As introduced in the previous section, the cooperation partner faces two situations when deciding about an act of cooperation: the situation of the bargaining and that of the task execution. In the first one, the cooperation partner negotiates about the task to perform and, in the end, claims to achieve a certain cooperation outcome. The other's statements and behaviour in this situation can indicate its attitude towards the act of cooperation and can thus correlate with the task execution. They are related to a belief about the other's *willingness*, which has been proposed by Falcone et al. (see Section 4.2). Consequently this situation features two independent entity types for an object-oriented data model: the *cooperation partner* and the *bargaining context*. They are related with one another in two types of relationships. One describes, how the partner *negotiates in* that situation independently of the task, and, thus, independently of the trustee's task-related ability. The other relationship type contains the final agreement about the cooperation outcome and the task-dependent negotiation characteristics. This it what the trustee *claims* to achieve. Figure 6.3 shows all these types on the left.

Later the cooperation partner performs the cooperation task. This situation features the behaviour of interest and requires certain abilities of the trusted system. So it is related to Falcone et al.'s belief about the other's *ability* or *competence* (see Section 4.2). This situation again includes two independent entity types: the *cooperation partner* and the *task context*. One of

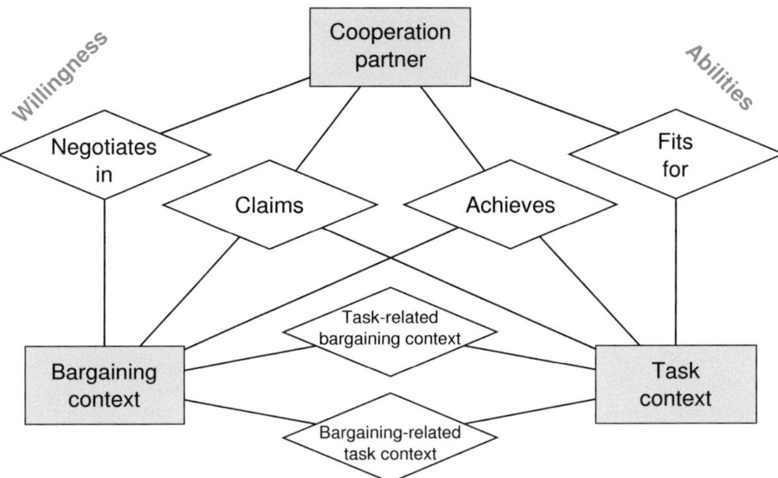

Figure 6.3: An entity-relationship context diagram describing the data included in the Enfident Model.

the relationship types between those two entity types characterises, how the trustee's ability *fits for* the task independently of the bargaining. Another relationship type describes the real cooperation outcome and the willingness-related behaviour during the task execution. This is what the trusted system *achieves*. Figure 6.3 shows these types on the right.

The bargaining context and the task context may also be in a relationship, although this relationship seems to be less important. To emphasise the temporal progress, the model contains two relationship types between those entity types: the *task-related bargaining context* and the *bargaining-related task context*.

All these types are root types. Sub-types can be defined, if a system acts in different settings. So a trusting system may be designed for various kinds of tasks and may cooperate with different kinds of cooperation partners. The main statement of this overview is that every trust situation is made of these kinds of entities and relationships, either explicitly or implicitly. A specific situation then consists of instances of those types. Finally note

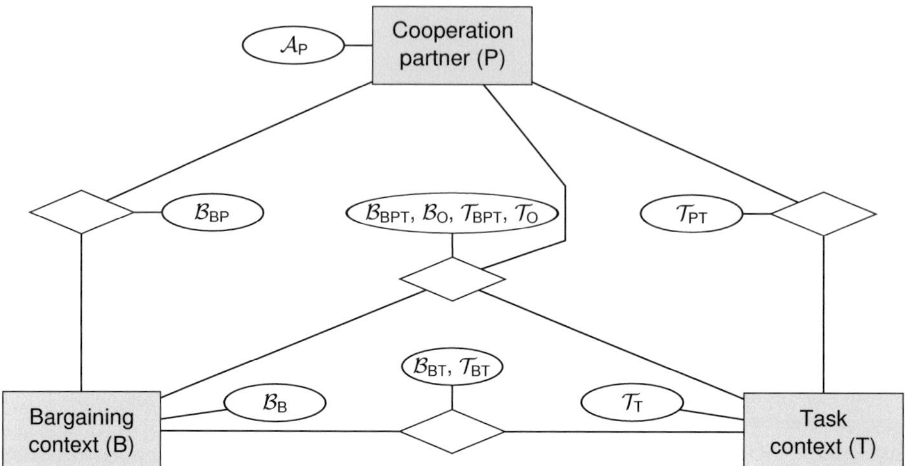

Figure 6.4: The entity-relationship diagram with the attributes. To maintain a clear arrangement, the relationships that connect the same entities are collapsed.

that the assignment of willingness and ability to some parts of the model is not meant to be exclusive. It should help to understand the model and derive algorithms for specific scenarios. Properties of the task context, for example, are expected to correlate stronger with ability than with willingness. But certainly correlations with willingness could also be found.

6.2.2 Detailed Description of the Entities and Relationships

In this subsection, the entities and relationships are detailed by describing their attributes. Figure 6.4 extends the Figure 6.3 by the attributes. Note that the Enfident Model makes no assumptions about the nature of an attribute. It may be continuous or discrete, one- or multi-dimensional. Throughout this section, let P, B and T be types, which are derived from the root types cooperation partner, bargaining context and task context, respectively.

The set of attribute types $\mathcal{A}_P$ describes the *cooperation partner* type P. In the logistics centre scenario, these attributes could, for example, be a robot's number of arms, its size, its weight and known defects. All these attribute types should characterise the cooperation partner independently of any specific situation. So they could also include a task-independent reputation of the partner or the time how long the partner is already working at the logistics site.

The *bargaining context* type B represents the situation of bargaining. Its set of attribute types is denoted by $\mathcal{B}_B$. For example, the presence of witnesses as well as the time and the location of the bargaining characterise the bargaining context. Furthermore in the vehicular network scenario, the current traffic situation could indicate the willingness to tell the truth. The attribute types in $\mathcal{B}_B$ should characterise the bargaining context independently of the cooperation partner. Sometimes there happened no bargaining though. The desired outcome and way of cooperation are implicitly assumed. In the Enfident Model this is still a certain form of bargaining.

The cooperation partner lives in the current bargaining context and may exploit it. So the types P and B are in a relationship, which is of the type *negotiates in* and which is independent of the cooperation task. Its set of attribute types is regarded by $\mathcal{B}_{BP}$. The relationship may reflect the progress of the bargaining. Did the partner, for example, change his opinion after a while? Happened the bargaining several times? Was the progress of the bargaining expectable? Was it quick? All in all, this relationship type models those aspects of the other's behaviour and communication that are independent of the task.

During the bargaining, the cooperation partner indicates what he is willing to do. So when defining the sets $\mathcal{B}_B$ and $\mathcal{B}_{BP}$, the system designer may think of indicators for *willingness* and self-confidence. Furthermore note that the willingness and thus its indicators could change over time. An implementation should consider this (see Section 6.3).

The set of attribute types $\mathcal{T}_T$ is associated with the *task context* type T. It characterises the cooperation task and the context of the task execution. Task attributes of the logistics centre scenario could be the forms of the containers, the source and target location, the tools necessary for the work, or the fact, whether the task is executed at night. All in all, these attribute types should characterise the task context independently of the cooperation partner.

The cooperation partner performs the task in its context. So the types P and T are in a relationship that is related to the other's abilities but independent of the bargaining. This relationship type is derived from the root type *fits for* and described by the set of attribute types $\mathcal{T}_{PT}$. In the logistics centre and the virtual market place scenario, these attribute types regard, for example, the experience of the cooperation partner: Do others report that the trusted system has done similar tasks before? Or does it offer many related products? So a task-related reputation of the cooperation partner would be part of this relationship, too.

The task execution demands the skills and abilities of the cooperation partner. These characteristics are learned by the trusting system from past interactions. They help to reason about the cooperation outcome. But even if the partner's abilities are unknown, the context of the task, the manifest properties of the trustee and the agreed cooperation outcome could be checked for their plausibility. So thinking about the other's *abilities* and *competence* as well as the *plausibility* of the agreement helps the system designer to find attributes for the sets $\mathcal{T}_T$ and $\mathcal{T}_{PT}$.

The situations of bargaining and task execution are connected through the trusting and the trusted system. The relationship types *task-related bargaining context* and *bargaining-related task context* reflect this. Relationships of these types are associated with the attribute types in the sets $\mathcal{B}_{BT}$ and $\mathcal{T}_{BT}$, respectively. For example, the attribute types in $\mathcal{B}_{BT}$ could describe the availability of other cooperation partners during the bargaining and their competence. In the virtual market place scenario, the availability could be understood as the relation of supply and demand for the target product at a specific market place. The cooperation partner considers the task during the negotiation. Thus the attribute types in $\mathcal{B}_{BT}$ might also be related to its willingness.

Finally the relation between all three entities remains. Here again, the model distinguishes the situations of the bargaining and the task execution with two relationship types. In the bargaining situation, the trustee *claims* to achieve a certain cooperation outcome. The agreed outcome is described by the set of attribute types $\mathcal{B}_O$. All other attribute types of this relation are collected in the set $\mathcal{B}_{BPT}$. Later the trustee performs the task and *achieves* a cooperation outcome. This actual outcome is described by the set of attribute types $\mathcal{T}_O$, while the set $\mathcal{T}_{BPT}$ contains all other attribute types of this relation.

The outcome attribute types in $\mathcal{B}_O$ and $\mathcal{T}_O$ assess all facets of the cooperation outcome. They usually describe the same aspects of the cooperation outcome, but can be realised with different values. Section 6.5 further discusses how the outcome can be rated. In the vehicular network scenario, the values in a received message, like the speed limit value, and their errors describe the cooperation outcome. In the logistics centre scenario, the duration of the task execution and the intactness of the transported goods are examples of outcome attributes. So the outcome can influence various attitudes towards the partner: competence, honesty, willingness and some more. This is discussed in Section 10.2.2.

The sets $\mathcal{B}_{BPT}$ and $\mathcal{T}_{BPT}$ contain the attributes, which are related to all three entities except for the outcome attributes. For example, these attributes could reflect that the cooperation partner claims he has done a similar task before, or that the same witnesses are present in both situations.

6.2.3 Conclusion

This section introduced the entity-relationship view on the Enfident Model. It is quite complex as it describes all involved data. In doing so, it fulfils Requirement 1 on page 77. The data is defined, on the one hand, by three entities and their relationships and, on the other hand, by attitudes about the trustee's competence and willingness. In this way, the Enfident Model unifies the attitude-based and the experienced-based view on trust in the literature. Section 10.2.2 further discusses the connection between both views and their relation to the Enfident Model.

The entity-relationship view of this section disregards the time. Of course, there are several acts of cooperation with different configurations of participating entities. They make up the experience of the trusting system. And certainly, the bargaining and the task execution happen in a sequential order. Some pieces of information can already be known, others could be unknown. This temporal view must be kept in mind. The presented entity-relationship model organises the characteristics of a trust situation at just one point in time. That means, every single situation in Figure 6.2 on page 99 is described by the proposed triplet of entities. In this way, the attributes may change from situation to situation.

So far, I described the data involved in reasoning about trust. But what can be done with the data? The next section shows how the pieces of data are connected with each other. And afterwards, I detail how the model produces some output.

6.3 Reasoning Process

As pointed out in Section 4.1, reasoning about trust is mainly possible, because several acts of cooperation are correlated. Hidden causes connect them. The attributes in the previous subsection are the observable description of the hidden causes. In the entity-relationship model, the attributes also describe the entities. Therefore the entities correspond to the hidden causes. There are several possible causes, also regarded as the cause components. An observable attribute depends on a subset of these components. An entity is considered to correspond to a superposition of cause components. In the model, the superposition is represented by a cause variable for every entity. These cause variables may be multi-dimensional variables and are indicated as grey ellipses in Figure 6.5 on page 108.

The cause variables are important for the trust model to realise the generalisation over all experiences, which was required in Section 4.4.2. If an entity is assigned to every cause component but at different degrees of membership, the model can generalise and specialise. A new, unknown entity can belong to all components with degrees of a general representative. The more an entity is known, the more specific the degrees become. Thus learning and inference happen across entities, but especially for well known and similar entities. This way, the trust model can give reasonable advice even for partly or widely unknown trust situations.

Up to here, the inner state of an entity is modelled as a specific superposition of cause components. However the way an entity influences the cooperation outcome may change over time as Section 4.4.3 points out in detail. Thus the cause components associated with every entity may change from interaction to interaction. This is especially true for the cooperation partners, but may also be relevant for the task and bargaining contexts. For this reason, a time-varying model for entity types is proposed in the following. It can be applied to all root types but also to some of them only. The time-varying model assigns sub-entities to every entity. They represent observations of the main entity. An observation is a set of observed attributes. Thus an observation entity is connected with all the attributes that were assigned to a main entity in the previous subsection. In addition, it has its own cause variable, which the attributes depend on. This new cause variable has most likely the same value as its temporal predecessor of the same entity, but it may have changed. So the new hidden cause depends on its predecessor. Figure 6.6 on page 109 visualises this. It features the observation entity types instead of the main entity types. The observations of the same main entity are indexed consecutively. The additional relationship types *Next* connect an observation with its successor. Every observation has a cause variable, which determines the associated attributes. The figure hides the main entities. They are only relevant to determine what observation entities are connected through a Next relationship.

The following consideration may help to realise the constraint of a cause variable on its predecessor. A change of an entity's inner state is more likely the more time has elapsed since the last act of cooperation. For example, when two robots meet again on the next day, they have the same inner state with a higher chance than in the case when they meet again only after ten years. So the state change likelihood should incorporate the time.

With this time-varying model, the effect of an entity on the cooperation outcome may change over time. But also the observable entity attributes can change over time without breaking the model. An obvious hardware defect, for example, could have happened to a partner. This fact leads to a changed attribute value for new observation entities but not for past ones. Altogether the time-varying model connects the temporal view of Section 6.1 and the entity-relationship model of Section 6.2. The observation of a cooperation partner or a situational context results in a set of attributes. It is

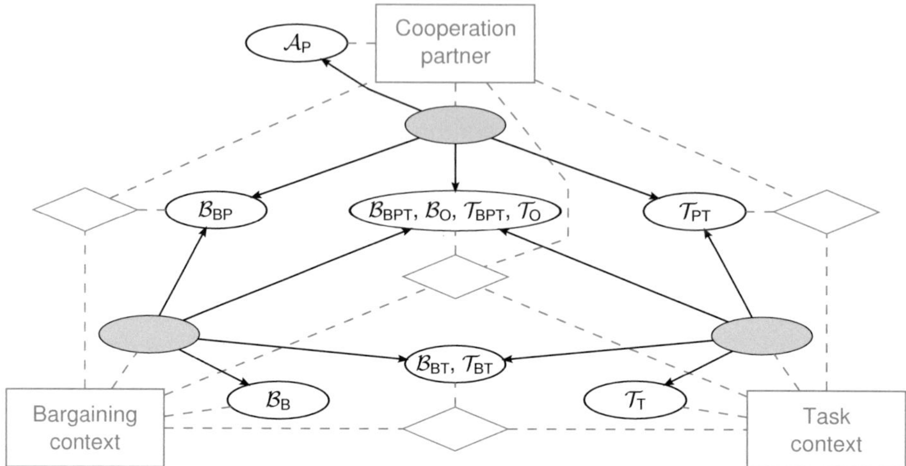

Figure 6.5: Reasoning with time-invariant entities. In this probabilistic entity-relationship diagram, the arrows describe the causality between the attribute types. In a realisation of the model, arrows to attributes types of relationship types are restricted to entities that participate in the relationship. (This is the implicit constraint that was mentioned in Section 5.2.)

a snapshot of the underlying causes that are in effect at that single point in time and represents a record with the form of the entity-relationship model. Such a record is repeated for every trust situation. The time-varying model connects these repetitions. Finally note, whether an entity type requires a time-invariant or a time-varying form of modelling, depends on the application.

With the setting of hidden causes and observable attributes described so far, the reasoning goes as follows. From the observed attributes, their causes can be inferred. Various techniques for probabilistic clustering like latent variable models (e.g. Bishop, 2007; Xu et al., 2009) come to mind. For time-varying entities, every cause variable is inferred with a constraint on its predecessor. Then with the cause variables, the unknown attributes can be estimated. Typically these unknown attributes are the attributes of the cooperation outcome that should be predicted. Note that some reasoning techniques perform those two phases in one step, others in two steps. And

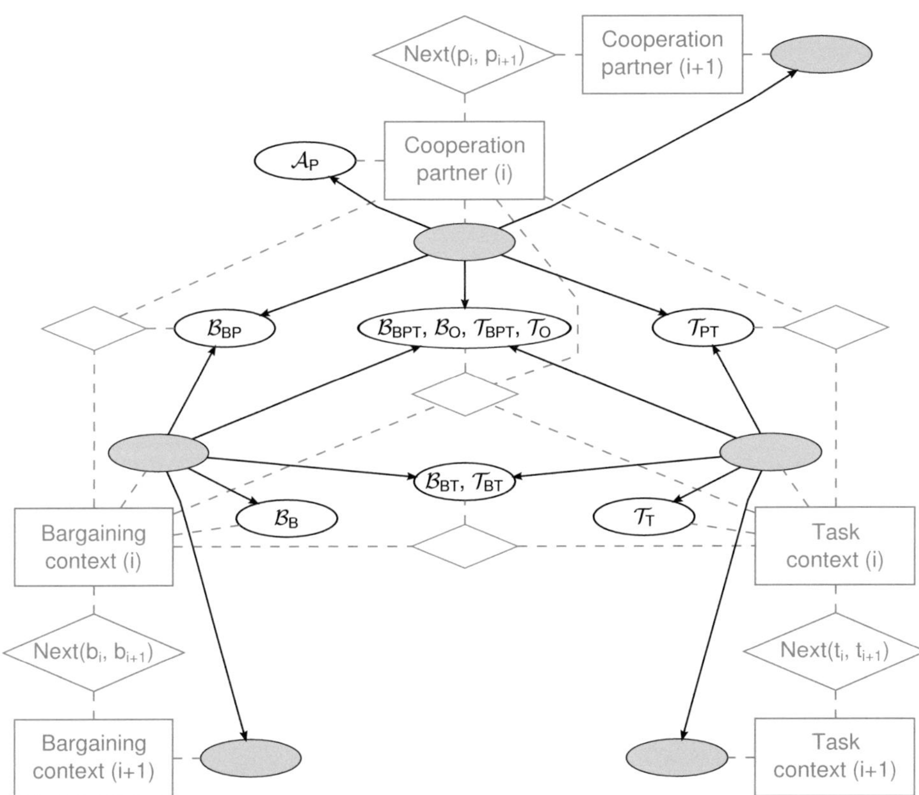

Figure 6.6: Modelling of the entities as time-varying processes. This is again a probabilistic entity-relationship diagram that depicts the causality in the model and, thus, the reasoning flow. This time, the entities represent observations. The observations of a main entity are number sequentially with the variable *i* in the graphic. It represents the progress of time. (Note that the variable *t* is already reserved for entities of the type task context.)

some make the hidden cause variables explicit, others only work with the known and unknown observable attributes.

In summary, the reasoning is supported by characteristics of each entity and by past cooperation outcomes (Requirement 4). These two information sources together with the hidden causes let the model generalise and specialise as needed for trust (Requirement 6). The cause variables make it also possible to capture temporal changes of entities (Requirement 7). And finally the Enfident Model needs not forget anything or recompute something, because it is just a conceptual model. For this reason, it inherently respects Requirement 5.

6.4 Querying

So far, the presented Enfident Model learns models of other system's cooperation behaviour. The structure and the parameters of a learned Enfident Model contain models for many of those other systems. Based on this data, a realisation of the Enfident Model can infer various reasonable values not just a specific output value. For this reason, I say that another module can query the Enfident Model for a value that describes a certain aspect of the other's cooperation behaviour. This section describes some typical forms of values that can be queried from the Enfident Model. Note that an implementation will probably have something like output values, which are pre-determined queries.

A trust model is typically asked to predict the outcome of a planned act of cooperation. Most informative would be the complete probability distribution for all outcome attribute types in O_A: a probabilistic trust distribution as introduced in Definition 2.1 on page 32. A decision module can then prioritise the attribute types and apply utility functions on the distributions. But it may also be that only a subset of the attributes like danger-related attributes are necessary to decide pro or contra cooperation in a certain situation. The information of interest could also be an interest-related trust distribution, which incorporates utility and risk as proposed in Definition 2.2.

While distributions over all possible values of an outcome attribute are most informative, trust values that summarise the content of the distribution are widely spread (see Section 4.3). The Enfident Model can be queried for such a measure, which is derived from trust distributions. Thus the querying mechanism is also the interface to some recommendation systems as Section 6.5 details. There is no single right way to derive a trust value from an outcome distribution. Various ways have been proposed, which emphasise different aspects of the outcome distribution as Section 4.3 shows. Some trust values represent the mean behaviour of the other system; others pay special attention to the negative behaviour. Again some others distinguish the positive and the negative behaviour with a trust and a distrust value. All these forms have their own right for certain application scenarios. Consequently the Enfident Model leaves it open, how a trust value is computed. Some appropriate algorithms can be found in related work (Section 4.3). Typically they include utility or loss functions and aggregation functions, which reduce the outcome distributions of several attributes to an overall trust value for the whole act of cooperation (see, e.g., Calvo et al., 2002).

So far, the outcome of a specific cooperation situation was predicted. Sometimes a more general assessment of the partner or of any other entity may be necessary. When talking about trust, for example, the trust in the partner disregarding the context could be relevant. And as a second example, the influence of each entity on the cooperation outcome must be investigated in order to decide what measures would be best to make a good outcome more likely. All these kinds of assessments can be queried from the Enfident Model. The reasoning happens then by inserting attributes and latent variables as needed for the query. For example, to get a general impression a certain cooperation partner disregarding the context, an act of cooperation with a specific partner entity and unspecific context entities must be inserted in the model. This means, the latent variable of the partner entity is described by all its attributes and past cooperation outcomes, while the other latent variables have completely unspecified attributes. Then the Enfident Model can predict probability distributions for the cooperation outcome and derive a trust value as needed.

While a probability distribution over all values of an attribute is quite informative, it does not quantify the certainty of the distribution. The same distribution could be obtained from one interaction or one hundred interactions with the cooperation partner. So the certainty or ignorance behind a prediction is of interest as well. One way to express the ignorance of the trust distribution is the amount of evidence the distribution has been calculated from (see, e.g., Jøsang et al., 2006). Since the Enfident Model learns across entity boundaries, an outcome value cannot be said to origin from exactly those interactions with the cooperation partner. But still, the number of experiences with a certain entity combination gives a good ignorance indicator. Another way to express the ignorance are the variances of the parameters of the provided probability distribution.

In summary, various outcome ratings can be obtained from the Enfident Model, as long as the corresponding ratings of previous outcomes have been included as attributes. Then a probability distribution of any attribute can be queried. This is possible for a specific act of cooperation with a well defined cooperation partner, task context and bargaining context. But it is also possible to obtain a more general assessment, if some of the attributes and entities remain unspecified. Instead of a probability distribution, trust values can also be received by plugging in the desired algorithms. Because the mapping from an outcome distribution to a trust value reduces information, I recommend the whole distribution over a trust value wherever possible.

6.5 Implementation Notes

Up to here, the Enfident Model has been introduced as a general framework to design a trust algorithm for a specific application. The description may be too abstract to implement a trust algorithm directly. Therefore this section provides an addendum of, how selected ideas from the literature relate to the Enfident Model. Some of them are out of the scope of this dissertation. I still mention them here to connect the Enfident Model to the surrounding research and to point to further interesting research areas.

6.5.1 Determining the Attributes

The various entities and relationships in the Enfident Model can be characterised by many attributes. Some of them may be more important for a certain application than others though. And some attributes might be more helpful for the reasoning algorithm to converge and predict than others. This subsection notes some examples of those attributes and considerations on how to use them.

Knowing that a trustee belongs to a certain group or category, may influence trust, even if no experiences with him are available. If a human partner, for example, is a doctor, he is more trustworthy regarding health advises than otherwise. So the group membership characterises the cause component of the trustee and is, thus, an attribute of the partner entity type.

When a third party has recommended the partner or discouraged from him, this rating corresponds to a task-related or general attribute of the partner entity. Recommendations should be taken with care like any other information from others. This means that trust in recommendations of a third party must be respected (e.g. Jøsang et al., 2006). Chapter 4 gives some references for recommendation and reputation systems.

The outcome attributes describe the behaviour of the cooperation partner. Therefore they are especially important for trust development. They should comprise *all facets* of the cooperation outcome. This way, the decision module can balance them as needed in the current situation. If the task, for example, is to cut a certain steel beam, the resulting size of the beam and the task duration could be chosen as outcome attributes. In this example, the real outcome can be represented in several ways. Should the absolute task duration be taken for $\mathcal{T}_O$ (e.g. 0.9 h)? Or should a rating with regard to the agreed duration be used (e.g. the deviation 0.1 h if the agreed duration was 1.0 h)? Or should a rating with regard to a desired ideal duration characterise the outcome (e.g. the deviation -0.4 h if the desired duration was 0.5 h)? All these values are different views on the same thing. What values suit better depends on the reasoning algorithm and the desired output of the trust model.

In the vehicular network scenario, a message could contain a statement together with the sender's own certainty about the statement. A vehicle could say that there is a new sign limiting the speed to 60 km/h, but it is not sure about this with a certainty level of 0.2? What is the outcome rating, if the sign actually shows a limit of 80 km/h? The sender already said, it is uncertain regarding the value. This example addresses the difference between the absolute correctness of the statement (an ability regarding sign detection) and the discounted correctness of the statement (an ability regarding telling the truth). Both are independent and should be included as two separate attributes. This becomes even clearer, if the number of the provided independent values is considered: The sender sent the two values 60 and 0.2, so two outcome properties can be evaluated. Bamberger et al. (2012) discuss this problem and also consider the case that the own observation, that is, the assumed true value is uncertain as well.

A few outcome attributes could be subject to norms. They are implicitly assumed by the cooperation partners without explicit bargaining. Some norms are enforced by law or site policies. They are more likely to be achieved. If the enforceability is varying (e.g. because of varying sites or countries), this fact should be included as an attribute. Other norms are mere conventions. In this case, both cooperation partners could follow different norms. Moreover norms could change between the time of bargaining and task execution. All in all, norms should also be modelled as outcome attributes, so their violation can be respected.

6.5.2 Distinguishing Entities

In some applications, it is possible to recognise a cooperation partner again and again, for example, with techniques from cryptology. But how can a task context be said to be the same as a previous task context? Basically two task or bargaining contexts can be taken as the same, if they have exactly the same attributes. But it is also valid to represent every task context as its own entity. The reasoning works with both ways of realisation. The same problem appears, if cooperation partners cannot be identified uniquely. Such a scenario can also be realised in both forms.

If the entities for task and bargaining contexts are reused, the relationships could be realised with different attributes. In the strict sense of the entity-relationship model, a kind of trial or situation counter must be incorporated as another entity to allow several realisations for a single relationship. I did not regard this in the text so far to keep the description clear. But certainly, it is possible to have this repetition of attributes for every trust situation.

6.5.3 Trusting a Group of Systems

When a system commissions a couple of other systems to reload some goods, the system could be said to trust this specific group of other systems. How can a group of systems be represented in the Enfident Model? In some cases, an aggregation function could be appropriate to transform the trust in the individual group members to the trust in the whole group. If the task, for example, can be split in independent sub-tasks, which are assigned to individual trustees, the whole cooperation outcome is a combination of all sub-task outcomes. Or the cooperation outcome could, for example, be determined by the weakest member of the group. All these aggregation functions depend on the specific application.

In other cases, the group can be represented as an entity on its own. Then its cause component can be learned as well. The group entity can have relationships to the entities of its members. These links describe the group entity like attributes of the types in $\mathcal{A}_P$. Such additional links are not in the scope of this dissertation. Further research is necessary on this subject.

6.5.4 Modelling the Social Structure

Some authors emphasise that the social structure around the trustor and the trustee is an important source for trust development (e.g. Buskens, 1998; Jøsang et al., 2006; Sabater and Sierra, 2002; Yu and Singh, 2002). They propose different ways to incorporate the social structure. If the trustee belongs to a certain group or clique, he can be characterised by a group membership attribute as described above. If third parties recommend the trustee, the recommendations can be used in the way discussed above too.

Moreover friendship-like connections become manifest in recommendations and possibly high cooperation frequencies. They can be modelled as additional relationships between the systems' entities together with an attribute for the relationship existence or relationship strength (Chu et al., 2007; Heckerman et al., 2007; Taskar et al., 2007). This way of modelling expresses that the cause components of connected entities correlate with each other. The references show promising results. How these techniques perform for trust development is subject to further research though and not in the scope of this dissertation.

6.5.5 Designing the Reasoning

Although the Figures 6.5 and 6.6 use the graphical notation of Bayesian networks, the Enfident Model leaves it open, what reasoning technique is used. The Bayesian network notation should just visualise the causal concept. An implementation may well apply techniques like fuzzy logic, Gaussian processes, Markov logic networks or some other methods for latent variable models. The evaluation in the paper realises the Enfident Model with Bayesian networks based on finite mixture models.

The reasoning technique should be able to realise the requirements of Section 4.4. Especially it should respect the generalisation requirement, because this requirement emphasises the purpose of trust to support the decision making in unknown situations.

Not all characteristics of an entity or relationship support the inference of the cause component. Such attributes could be attached with an additional probabilistic or deterministic logic. For example, the difference between the real outcome and the agreed outcome could, in some cases, better support the inference of the cause component than the absolute value of the real outcome. Then this difference should be chosen as the child node of the cause variable.

6.5.6 Connecting the Enfident Model to a Reputation System

Reputation systems provide good hints, how trustworthy a cooperation partner might be, especially if only few own experiences are available (see Chapter 4 for references). These hints can be incorporated in the Enfident Model as additional attributes. For example, if the reputation system provides one reputation indicator for every participating system, this indicator would be an attribute of the cooperation partner. In contrast, if the reputation system is task-aware by providing a reputation indicator for every combination of task and participating system, this indicator would be an attribute of the task-partner relationship. If there are many reputation sources, they should be included as separate attributes with separate parameter sets (e.g. separate conditional probability distributions). This way, the reasoning process can learn and distinguish the reliabilities of the different reputations sources. This is similar to the recommendations mentioned in Section 6.5.1.

In the other direction, an agent must feed a reputation system with its experiences. Usually the uploaded data only reflects the rating of one certain cooperation outcome. No prediction, and thus, no trust is involved in this process. Moreover, these computations are specific to the policies of the reputation system. So a distinct reputation module would compute these ratings and pass them to the reputation system. This processing is not related to the Enfident Model. In contrast, some other reputation or recommendation systems exchange trust values that originate from all past interactions with the partner. Such trust values can be queried from the Enfident Model as described in Section 6.4.

The implementation notes of this section still address the conceptual level. The next chapter switches to the algorithmic level. It proposes a probabilistic schema from statistical relational learning to realise the Enfident Model.

7 Reasoning Algorithms for the Enfident Model

The probabilistic entity-relationship model of the previous chapter does not prescribe a single way of realisation. This chapter proposes an exemplary implementation, which is later used for the evaluation of the Enfident Model. It does not aim to be the best implementation following any benchmark. But it is an implementation that demonstratively explains the Enfident Model. It stays close to the ideas that underlie the Enfident Model, although that might impose some algorithmic disadvantages. This is appropriate, because Chapters 8 and 9 evaluate the Enfident Model itself and do not compare several algorithms for the Enfident Model.

Although this chapter targets at the Enfident Model, it presents a generic way to transform a probabilistic entity-relationship model into a specific probabilistic model and shows, how to infer in this model with Gibbs sampling. As a consequence, the proposed technique can be applied to any other probabilistic entity-relationship model as well.

The work in this chapter is based on that of Xu (2007). It extends her concept, in Sections 7.2.2 and 7.3.2, by allowing a hierarchy of entity and relationship types and, in Sections 7.2.3 and 7.3.3, by extending entities to have time-varying states. Sections 7.2.3 and 7.3.3 are also based on relational dynamic Bayesian networks (e.g. Manfredotti, 2009) and non-parametric hidden Markov models (Van Gael, 2011). This chapter recalls some common techniques for Bayesian inference to relate them to trust and to describe all algorithms in a consistent notation with type hierarchies. The theoretical background on the algorithms is beyond the focus of this thesis though. Please consult the three mentioned dissertations for this. They describe well the state of the art at their time.

7.1 Introduction

Section 4.1 argued that the observable attributes of a trust situation depend on hidden causes. This idea corresponds well with the mixture models in statistics. I use the Bayesian approach to mixture models with prior distributions on the parameters, because in this way, the parameters can easily be specialised for objects and types depending on requirements. The reader can find an introduction to Bayesian mixture models in Bishop, 2007, for example.

In a mixture model, an observation A comes from a superposition of several independent effects. Every effect is modelled with a probability distribution. The overall probability distribution, the *mixture distribution F^M*, is a weighted sum of all sub-distributions, the *mixture components F_k*. In a finite mixture model, the number K of independent effects and, thus, of mixture components is fixed and finite:

$$A \sim F^M = \sum_{k=1}^{K} \pi_k \, F_k. \tag{7.1}$$

The variables π_k are the *mixture weights*.

However the main idea of Section 4.1 is that several observable attributes are correlated, because they share common hidden causes. As a consequence, for the object-oriented setting considered here, the modelling concept above must be extended. Several mixture models must be entangled. This is done by introducing a random variable Z, which connects the correlated attributes. It indicates which mixture component is selected with which weight and is called the *indicator variable* or *component variable*. The values of this variable are taken from a set S with K elements. Z follows a categorical distribution with the probability vector $\pi = (\pi_1, \ldots, \pi_K)$. (The categorical distribution matches the multinomial distribution if the number of experiences $n = 1$; it is the distribution of each trial in a multi-valued urn experiment.) A Bayesian mixture model realises the vector π of mixture weights as a random variable Π, which follows a Dirichlet distribution with the parameter vector α. Altogether the complete probabilistic model of a

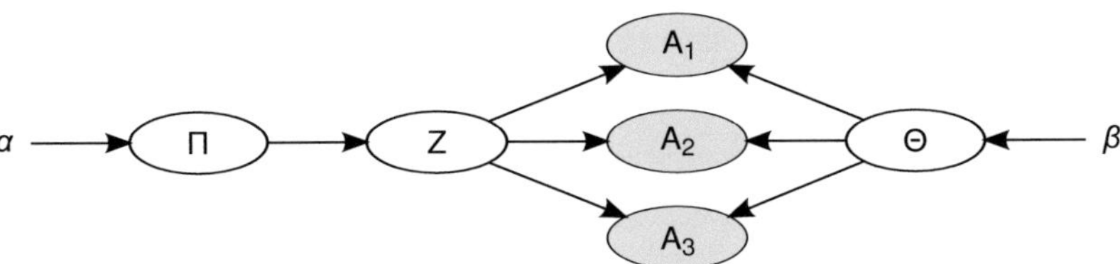

Figure 7.1: Exemplary Bayesian network of a Bayesian mixture model. A_1, A_2 and A_3 are three observations of the same statistically process, which is described with the mixture model. Θ is a tuple of all component parameters Θ_s with $s \in S$.

Bayesian mixture model with indicator variables goes as follows:

$$A \mid Z = s \sim F_s$$
$$Z \mid \Pi = \pi \sim \text{Categorical}(\pi)$$
$$\Pi \sim \text{Dirichlet}(\alpha)$$

To simplify the notation in the above equation, the set S is assumed to be $\{1, \dots, K\}$ without restricting it to this content. That way, $s \in S$ can serve as an index for F. Further note that the probability distribution for A becomes the same as in Equation 7.1, when taking π as a fixed parameter and integrating out Z.

In the context of Bayesian inference, the mixture components have all the same functional form $F(\theta)$. Then each component is represented by a parameter vector θ_s with $s \in S$. Through the subscript s, a component is associated with the value s of the state variable Z. All vectors θ_s can in turn be understood as random variables Θ_s and associated with a prior distribution. This prior is a distribution over distributions. Figure 7.1 illustrates the setting in form of a Bayesian network.

Note that the mixture model with an indicator variable still assumes that there are exactly K causes and that these are statistically independent. However the additional component variable has the advantage that it can be interpreted as a random variable for hidden causes behind the observable attributes. This interpretation matches with an assumption of the Enfident Model: There are hidden causes behind a task context, which affect the cooperation outcome; and the behaviour of the cooperation partner depends on its inner states. To underline this view, I additionally denote the indicator variables as *cause variables* or *state variables*.

The remaining sections of this chapter show, how mixture models can be applied to an object-relational data set. I start with a finite number of components K and propose probabilistic models for objects with time-invariant and time-varying states. After that, I extend these models to the infinite-dimensional case.

> Initialise all variables $x_1, \dots, x_n$ randomly
> **Repeat** T *times* **do**
>   **foreach** $i \in \{1, \dots, n\}$ **do**
>   Sample x_i from $P(X_i \mid x_{-i}) = P(X_i \mid \mathrm{mb}(X_i))$.

Algorithm 7.1: Basic steps of the Gibbs sampler. It samples all random variables $X_1, \dots, X_n$ T times from their Markov blanket. $x_1, \dots, x_n$ contain the last sampled value of their corresponding random variable.

The common base for all the algorithms in these sections is the Gibbs sampling. This sampling technique makes it possible to combine different probabilistic models for the various entity types in a simple plug-in manner. The Gibbs sampling is widely spread and detailed, for example, in Bishop, 2007 and Andrieu et al., 2003. For the explanations in this section, it is sufficient to know that the Gibbs sampling of a Bayesian network with the unknown random variables $X_1, \dots, X_n$ basically involves the steps shown in Algorithm 7.1. Here the variable x_i denotes the newest value of X_i during the sampling and x_{-i} is the joint event of all random variables except for X_i in their current state. Because X_i depends only on the variables of its Markov blanket $\mathrm{mb}(X_i)$, it is sufficient to consider $P(X_i \mid \mathrm{mb}(X_i))$. The sampling goes over T rounds. This sketch of the algorithm shows that, first, $P(X_i \mid x_{-i})$ must be found for every node in the Enfident Model and, second, the algorithm, that is, the conditional distribution for one node can be different and is independent from that for other nodes. The following sections show ways to statistically model the entities in the Enfident Model and propose corresponding algorithms for $P(X_i \mid x_{-i})$.

7.2 Realisations with Finite Mixture Models

This section considers mixture models with a state space that has a finite and fixed number of dimensions K. Such a mixture model assigns the data probabilistically to K clusters. The corresponding sampling algorithms are easy to implement and lead in many publications to good results. They also provide a good basis to understand the more complex case of an infinite dimensional state space in Section 7.3. On the downside, in some practical problems, the number of necessary states changes over the lifetime of the system or can hardly be guessed in advance. Then the techniques of Section 7.3 could help.

The state variables in this section have a categorical distribution. The conjugate prior distribution for its parameter vector is the Dirichlet distribution. While I assume the distribution to be known, I give its definition in the following to clarify my notation. (A good tutorial is Frigyik et al., 2010.) After that a probabilistic model for entities with a time-invariant state is proposed as well as one for entities with a time-varying state.

7.2.1 The Dirichlet Distribution

The definition of the Dirichlet distribution uses an integral that is called the multidimensional beta function. It is introduced first.

Definition 7.1 (Multidimensional beta function). Let θ regard the vector $(\theta_1, \dots, \theta_n) \in \mathbb{R}^n$ and $\mathcal{S}$ be the standard simplex in $\mathbb{R}^n$: $\mathcal{S} = \{\theta \in \mathbb{R}^n \mid \sum_{k=1}^n \theta_k = 1$ and $\theta_k \geq 0$ for $k = 1, \dots, n\}$. ($\mathcal{S}$ is an $(n-1)$-dimensional surface in $\mathbb{R}^n$.) Let $\Gamma(z)$ denote the gamma function. Then for $\mathrm{Re}(\alpha_k) > 0$ $(k = 1, \dots, n)$,

$$B(\alpha) = \int_{\mathcal{S}} \prod_{k=1}^n \theta_k^{\alpha_k - 1}\, d\theta = \frac{\Gamma(\alpha_1)\,\Gamma(\alpha_2)\,\dots\,\Gamma(\alpha_n)}{\Gamma(\alpha_1 + \dots + \alpha_n)}.$$

The integral on the right is the multidimensional beta integral. When understanding the integral as a function $B(\alpha)$ with the independent variable

$\alpha = (\alpha_1, \dots, \alpha_n)$, this function is called the *multidimensional* or *multinomial beta function* (Andrews et al., 1999, Section 1.8).

The beta integral normalises the density of the Dirichlet distribution, as the following definition shows.

Definition 7.2 (Dirichlet distribution). For $\alpha = (\alpha_1, \dots, \alpha_n) \in \mathbb{R}^n$ with $\alpha_i > 0$ for all $i = 1, \dots, n$, the function

$$f(\theta_1, \dots, \theta_{n-1}) = \begin{cases} \frac{1}{B(\alpha)} \prod_{i=1}^{n} \theta_i^{\alpha_i - 1} & \text{for } \theta_1, \dots, \theta_{n-1} > 0, \theta_1 + \dots + \theta_{n-1} < 1 \\ 0 & \text{otherwise} \end{cases}$$

with $\theta_n = 1 - \sum_{k=1}^{n-1} \theta_k$ is a density function. Its corresponding distribution is called the *Dirichlet distribution* of order n (Frigyik et al., 2010; Schmidt, 2009).

So the Dirichlet distribution of order n is a continuous multivariate distribution over an $n - 1$-dimensional space. Because it has only support on the standard simplex and vanishes otherwise, it suits well to model the probabilities over a partition of n events. Note that there are also other distributions on the standard simplex. None of them needs to be more correct or appropriate than the others. People usually take the Dirichlet distribution, just because it is the conjugate prior of the categorical distribution and, thus, leads to reasonable algorithms.

7.2.2 Time-Invariant Entity Types

This section considers the case that an entity is all the time in the same state. In addition, the possible number of states an entity can be in is finite.

Probabilistic model for the state variables. Section 7.1 shows, how observable attributes can result from unobserved effects. In the mixture model, these effects are assumed to be independent and associated with a state variable. From an object-oriented point of view, an entity perceives its environment and reacts depending on its inner way of working and its inner

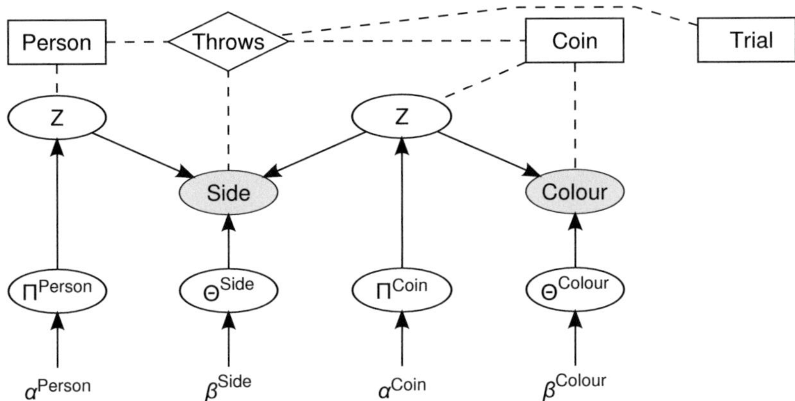

Figure 7.2: Probabilistic entity-relationship model of the coin toss example with finite time-invariant states. In addition to Figure 5.2 on page 93, the person that throws the coin is modelled, a state variable for every entity is added and the prior and hyper-parameters are made explicit. The parameter variables should not be repeated for every entity, so they are not connected to an entity type box.

state. Thus every entity is associated with a state variable. The possible states for an entity depend on its type. Two entities of the same type can potentially be in the same state. So the state variable belongs to an entity, while the set of states is associated with a root type.

For example, consider the coin experiment of Section 5.2. It features coins, which are thrown in trials. Here the model is extended by the persons, who throw the coins. One person may play fair, others may cheat. Then the colour and the result of a coin toss can be understood as observable hints about the hidden kind of the coin and the inner state of the throwing person. In the example, the kinds of coins are represented by a fixed set; and every coin is of one particular kind, which remains the same all the time. Figure 7.2 illustrates this setting. It is further explained in the following.

Let R be a root type whose entities should be modelled with a time-invariant state. $\mathcal{E}^R$ denotes the set of all entities that are of a type derived from R. So it contains all entities of interest in this section. Let the set of states S^R be a fixed finite set with K elements. I assume $S^R = \{1, \dots, K\}$

here without restricting it to that. This specific content of the set makes it possible to omit an additional index set for S^R at some places below. With these prerequisites, the entities in $\mathcal{E}^R$ are transformed into a probabilistic model in the following way.

The inner state of each entity $e \in \mathcal{E}^R$ is represented by a random variable Z_e with the image S^R. Z_e follows a categorical distribution with the parameter vector Π^R. The elements of this vector are the mixture weights. Following the Bayesian modelling approach, Π^R is taken as a random variable with a Dirichlet distribution. The parameters of this prior distribution are regarded by α^R. In summary, the probabilistic model of the hidden causes features a hierarchy of two levels of random variables as depicted in the expressions

$$Z_e \mid \Pi^R = \pi \sim \text{Categorical}(\pi) \quad \text{for all } e \in \mathcal{E}^R \text{ and}$$
$$\Pi^R \sim \text{Dirichlet}\left(\alpha^R\right).$$

The variable π is the current value of Π^R during the sampling. Figure 7.2 illustrates this model on the coin example.

Probabilistic model for the attributes. The inner state of an entity becomes manifest in its attributes. Thus all attributes of an entity are random variables that depend on the entity's state variable. In the same way, attributes of a relationship reflect the inner state of all participating entities together. Therefore a relationship attribute is a random variable that depends on the joint event of the indicator variables that belong to the associated entities. This results in the following probabilistic model for attributes.

An attribute is a random variable Y. It depends on the joint value s_Y of the indicator variables that belong to its associated entities. Let $S^{\text{type}(Y)}$ be the set of possible joint values. $\left(S^{\text{type}(Y)} = \prod_{e \in \text{entities}(Y)} S^{\text{roottype}(e)}\right.$, if $\prod$ denotes the Cartesian product.$\left.\right)$ s_Y is an element of this set. The attribute Y follows a mixture distribution with $|S^{\text{type}(Y)}|$ components. Every component has the form $F^{\text{type}(Y)}(\theta)$. It is identified by its parameter vector $\Theta_s^{\text{type}(Y)}$, which is associated with elements of $S^{\text{type}(Y)}$ through the subscript $s \in S^{\text{type}(Y)}$. In summary, the probabilistic model of the attribute Y has the following form.

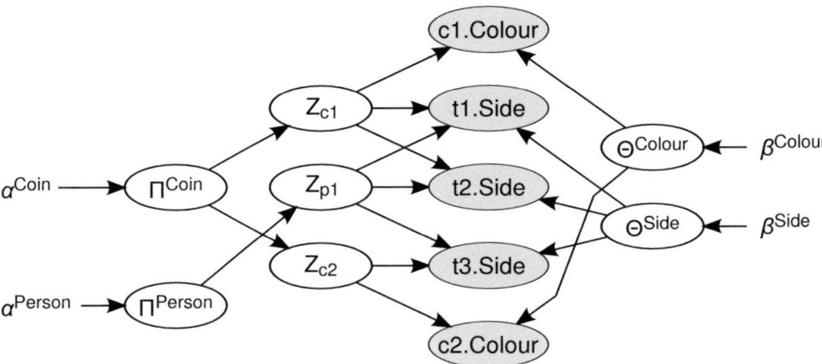

Figure 7.3: Exemplary Bayesian network for the probabilistic model with finite time-invariant states. The figure depicts, how several mixture models as that of Figure 7.1 on page 121 are entangled through a hierarchy of random variables. This hierarchy makes it possible to learn on all levels and transfer knowledge between otherwise separate probability distributions. c1 and c2 are entities of type Coin and p1 is an entity of type Person. Z_{c1}, Z_{c2} and Z_{p1} correspond to the entity attributes c1.Z, c2.Z and p1.Z, respectively. I aligned their names with the mathematical notation of the main text. t1, t2 and t3 are relationships.

Let $\theta_s^{\text{type}(Y)}$ be the current value of $\Theta_s^{\text{type}(Y)}$ and z_e the current value of Z_e. Then the attribute Y has the conditional distribution

$$Y \mid \{Z_e = z_e\}_{e \in \text{entities}(Y)}, \{\Theta_s^{\text{type}(Y)} = \theta_s^{\text{type}(Y)}\}_{s \in S^{\text{type}(Y)}} \sim F^{\text{type}(Y)}(\theta_{s_Y}^{\text{type}(Y)}).$$

Associated with that distribution is the likelihood function $L^{\text{type}(Y)}$. It provides the likelihood of the current value of Y given the parameter vector $\theta_{s_Y}^{\text{type}(Y)}$. The likelihood function is necessary below for the sampling algorithm.

To complete the model, the parameter vectors $\Theta_s^{\text{type}(Y)}$ can in turn be understood as random variables with a prior distribution when following the Bayesian paradigm. The detailed model of the attributes and their prior distributions is out of the scope of this dissertation though. It depends on the application and is not in the core of trust modelling.

Summary of the probabilistic model. The probabilistic model of this section can be summarised as follows. The state variables are related to entities (e.g. specific cooperation partners), while the parameter variables α^R and

Π^R are associated with root types (the types cooperation partner, bargaining context and task context). In the consequence, the a priori probabilities of the state variable depend on the type, while the a posteriori probabilities are individual for each entity and formed through the observation of several attributes. Figure 7.3 on the preceding page shows an exemplary realisation of the probabilistic model. It even better points at the hierarchy of the variables. Only the grey elements are observable. All others are hypothetical constructs. They make it possible to perform probabilistic clustering and learning at the same time, just by inference in the full Bayesian model. The state variables cluster similar entities, that is, similar groups of observations. Thus similar cooperation partners and similar task contexts, for example, are treated in a similar way. Every entity belongs to all clusters at the same time but with different weights. The parameter variables in $\Theta^\tau = \{\Theta_s^\tau\}_{s \in S^\tau}$ (τ is an attribute type) are learned based on the clustering. So basically, all evidence influences them, but weighted by cluster assignment probabilities.

Below I show, how Z_e can be sampled during Gibbs sampling (Algorithm 7.1 on page 122). Π^R is superfluous for this. It can be integrated out resulting in a compound probability distribution that depends on α^R. The sampling of Y is out of the scope of this dissertation, because it depends on the application-specific attributes.

Sampling with the parameter Π^R. For Gibbs sampling, a cause variable Z_e is sampled depending on its Markov blanket $mb(Z_e)$. Here the Markov blanket refers to four objects (compare this with Figure 7.3): The parents of Z_e are the mixture weights in Π^R. Thus the a priori probability for the state $s \in S^R$ is simply the value of Π_s^R. The children of Z_e are all attributes that belong to the entity e or to a relationship that e is participating in. The set of these attributes is denoted by A_e. The other parents of the children (except for Z_e) are the parameter vectors of the mixture components and, for relationship attributes, the cause variables of the other entities that participate in the relationship. Consequently the posterior update factor is the combined likelihood of all attributes in A_e depending on the current state of their parents. The likelihood function L^Y for the attribute Y was already mentioned above.

To take all this together in a compact equation, few more symbols are necessary for the values of the random variables. Given an attribute Y, let a_Y denote the current value of the attribute during the Gibbs sampling and s_Y the joint value of all state variables that Y depends on. s_Y selects the mixture component for Y. Let $\theta_{s_Y}^{type(Y)}$ be the current value of the parameter vector of that component. Furthermore the vector π denotes the current value of Π^R. Then the sampling distribution for Gibbs sampling is a categorical distribution over all states $s \in S^R$ with the probabilities

$$P\big(Z_e = s \mid \mathrm{mb}(Z_e)\big) \propto \pi_s \prod_{Y \in A_e} L^{\mathrm{type}(Y)}\big(a_Y, \theta_{s_Y}^{\mathrm{type}(Y)}\big). \tag{7.2}$$

The symbol $\propto$ means that the left and the right side are equal except for a certain factor, which is the same for all s. So the list of the values computed from the right side for all s must be normalised to sum up to one. This last step results in the distribution parameters.

Sampling with the collapsed Dirichlet-categorical distribution. The mixture weights can be integrated out to speed up the random walk of the Markov chain Monte Carlo simulation (see Neal, 2000, for example). Up to here, the vector of the mixture weights is a common parent variable of all considered state variables. Thus it makes the state variables independent from one another. As a consequence, when removing it, the prior probability of Z_e depends on all other state variables of the same root type. The resulting independence structure cannot be visualised as a Bayesian network any more, because all the state variables are mutually dependent now. The resulting algorithm still remains simple though, as all other variables except for Z_e are assumed to be fixed during the sampling of Z_e. This means that all other state variables are parents of Z_e just for this sampling step. I regard them by the set $Z_{-e}^R = \{Z_{e'} \mid e' \in \mathcal{E}^R \wedge e' \neq e\}$ in the following.

To avoid that the notation becomes confusing when deducing the algorithm without the mixture weights, I concentrate on one root type. So I can

remove the superscripts from α^R, S^R and $\mathcal{Z}^R_{-e}$. π is again a value of Π^R.
The aim is to solve

$$P(Z_e = s \mid z_{-e}, \alpha) = \frac{P(Z_e = s, z_{-e} \mid \alpha)}{P(z_{-e} \mid \alpha)}. \tag{7.3}$$

Here z_{-e} is the event that describes the current joint realisation of all state
variables $Z_{e'}$ except for Z_e ($Z_{e'} \in \mathcal{Z}_{-e}$).

The denominator can be got by integrating over the simplex $\mathcal{S}$ that is
associated with π. This results in

$$P(z_{-e} \mid \alpha) = \int_{\mathcal{S}} P(z_{-e} \mid \pi)\, p(\pi \mid \alpha)\, d\pi$$
$$= \frac{1}{B(\alpha)} \int_{\mathcal{S}} \prod_{s' \in S} \pi_{s'}^{n_{s'}}\, \pi_{s'}^{\alpha_{s'}-1}\, d\pi = \frac{B(\alpha + n)}{B(\alpha)}.$$

The function p regards the density function of the Dirichlet distribution; $n_{s'}$
is the number of all state variables $Z_{e'} \in \mathcal{Z}_{-e}$ that have the value s'; and n
denotes the vector of all $n_{s'}$.

The numerator of Equation 7.3 is the same as the denominator except
that the value of Z_e is counted. The vector of counts that includes Z_e is
denoted by $n^{(+Z_e)}$. So the prior probability of $Z_e = s$ is

$$P(Z_e = s \mid z_{-e}, \alpha) = \frac{B(\alpha + n^{(+Z_e)})}{B(\alpha + n)} = \frac{\Gamma(\alpha_0 + n_0)}{\Gamma(\alpha_0 + n_0 + 1)} \frac{\Gamma(\alpha_s + n_s + 1)}{\Gamma(\alpha_s + n_s)}$$
$$= \frac{\alpha_s + n_s}{\alpha_0 + n_0} \propto \alpha_s + n_s. \tag{7.4}$$

n_0 and α_0 are the sums over all $n_{s'}$ and $\alpha_{s'}$, respectively (with $s' \in S$). To
get the final result without gamma functions, the recursion $\Gamma(x + 1) = x\,\Gamma(x)$
helps.

The above result is very simple. It substitutes the probabilities π_s in Equation 7.2 on the preceding page. Together with the posterior update, the
probability of the state s depending on the Markov blanket is

$$P(Z_e = s \mid \mathrm{mb}(Z_e)) \propto (\alpha_s^R + n_s^R) \prod_{Y \in A_e} L^{\mathrm{type}(Y)}(a_Y, \theta_{sy}^{\mathrm{type}(Y)}). \tag{7.5}$$

$$p \leftarrow \alpha^R$$

for $s \in S^R$ **do**
    $c \leftarrow$ count the occurrence of s in all variables $z_{e'} \in \mathcal{Z}^R_{-e}$
    $p_s \leftarrow p_s + c$

    // Set the current variable to the correct state for the likelihood computation.
    $z_e \leftarrow s$
    for $Y \in A_e$ **do**
        $a_Y \leftarrow$ get the current value of Y
        $s_Y \leftarrow$ joint value of Y's parent variables
        $\theta \leftarrow$ select the component s_Y from the vector $\theta^{type(Y)}$
        $p_s \leftarrow p_s \cdot L^{type(Y)}(a_Y, \theta)$

$p \leftarrow$ normalise the vector p to sum up to 1
$z_e \leftarrow$ draw from Categorical(p)

Algorithm 7.2: Sampling of a finite time-invariant entity state. Here z_e is the variable of the state to sample (the value of Z_e). p is the probability vector of the distribution to sample from. The algorithm sets z_e in a new state that is sampled from its Markov blanket.

In this equation, the state variables are related to entities, while the parameter variables α^R and θ^τ, the counts n^R as well as the likelihood functions L^τ are associated with types.

The probabilities computed with the above equation are the parameters of a categorical distribution. During the Gibbs sampling, Z_e must be sampled from this distribution. Algorithm 7.2 summarises the equation in a programming style.

This subsection presented a probabilistic model, which assumes that the state of an entity remains stable over time. All observable behaviour is caused by the same system state. Regarding the state concept and the cause effect considerations, the probabilistic model is very similar to the considerations that have been made for the Enfident Model. This is the reason, why I selected it for this thesis.

7.2.3 Time-Varying Entity Types

The probabilistic model of this subsection allows the internal state of the observed system to change over time. In the consequence, every observation is a snapshot of the system's current state. The transitions of the system from state to state are assumed to be correlated. These correlations are represented by a Markov chain with state transition probabilities.

Probabilistic model. As an example, consider a coin, which can change its colour and its centre of mass electro-mechanically. (Do not think too much about the coin's shape, which would be necessary for this.) A simple state machine controls the transition from one colour to another and from one centre of mass to another. Then the colour and the result of a coin toss make up an observable snapshot of the inner state flow. Moreover imagine that the person, who throws the coin, sometimes intends to play fair and sometimes intends to cheat. Figure 7.4 visualises this example.

Let R be a root type whose entities should be modelled with a time-varying state. Again $\mathcal{E}^R$ denotes the set of all entities that are of a type derived from R. These are all the entities of interest in this section. The set of states S^R is pre-defined and finite. Every entity $e \in \mathcal{E}^R$ has been captured in T_e observations, which are numbered in their temporal order from 1 to T_e. The entities in $\mathcal{E}^R$ can be transformed into a probabilistic model in the following way.

For every observation $i = 1, 2, \ldots, T_e$ of the entity $e \in \mathcal{E}^R$, add a random variable $Z_{e,i}$ that represents the inner state of the entity during the observation. $Z_{e,i}$ with $i \neq 1$ depends only on its predecessor $Z_{e,i-1}$ (Markov property). It follows a mixture of categorical distributions. Given the event $Z_{e,i-1} = s$ with $s \in S^R$, $Z_{e,i}$ follows a categorical distribution with the parameter vector Π_s^R. Thus $Z_{e,i}$ has a mixture distribution. Π_s^R is modelled as a random variable, which has a Dirichlet prior distribution with the parameter vector α^R. All mixture components have the same hyper-parameters. The first state $Z_{e,1}$ has no predecessor. It simply follows a categorical distribution with the parameter vector Π'^R. This vector is again a random variable, which has a Dirichlet prior distribution with the parameter vector α^R. Fig-

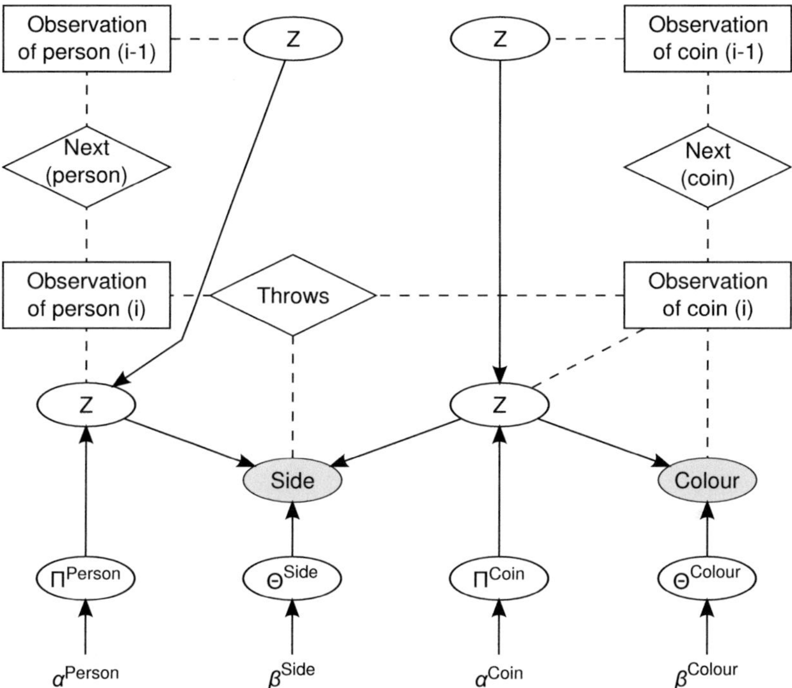

Figure 7.4: Probabilistic entity-relationship diagram of the coin example with finite time-varying states. Compared to Figure 7.2 on page 125, observations of real world objects are modelled instead of the objects itself. The observations of one object are still connected through the relationship next though. To keep the figure clear, I omitted Π'^{Person} and Π'^{Colour}. They can be imagined as being part of Π^{Person} and Π^{Colour}, respectively.

ure 7.4 illustrates the model using the coin example. The following mathematical expressions summarise the resulting probabilistic model of the state variables:

$$Z_{e,1} \mid \Pi'^R = \pi' \sim \text{Categorical}(\pi'),$$
$$\Pi'^R \sim \text{Dirichlet}(\alpha^R),$$
$$Z_{e,i} \mid Z_{e,i-1} = s, \Pi^R = \pi \sim \text{Categorical}(\pi_s) \text{ and}$$
$$\Pi_s^R \sim \text{Dirichlet}(\alpha^R)$$

for all $i \in \{2, \dots, T_e\}$ and all $e \in \mathcal{E}^R$. Π^R denotes the tuple of all Π_s^R ($s \in S^R$). π', π and π_s are values of Π'^R, Π^R and Π_s^R, respectively.

As in the previous section, the model features a hierarchy of random variable levels, with the observable attributes, the hidden states, the parameters and the hyper-parameters. The exemplary realisation in Figure 7.5 shows the hierarchy from left to right as well as temporal sequences of connected state variables from top to bottom. Again similar observations are clustered with the help of the state variables. The state transition probabilities are learned for all entities of the same root type, because they all use the same random variable Π^R. And if the parameter α^R is estimated with another hyper-distribution, it can incorporate the whole evidence of every mixture component in Π^R and transfer the knowledge about one mixture component to the others and to Π' (as, e.g., in MacKay and Peto, 1995).

Sampling of a state variable. In the following, I consider a state variable $Z_{e,i}$ that has a predecessor and a successor ($Z_{e,i-1} \rightarrow Z_{e,i} \rightarrow Z_{e,i+1}$). I regard the values of these variables with r, s and t, respectively ($r, s, t \in S^R$). $Z_{e,i}$ follows a mixture model. This means that the preceding state determines the mixture component of the prior distribution. It is the indicator variable of that mixture model. Given the preceding state, the setting is the same as in the previous section, except for the additional child $Z_{e,i+1}$. As a consequence, the prior probabilities in Equation 7.2 on page 129 must

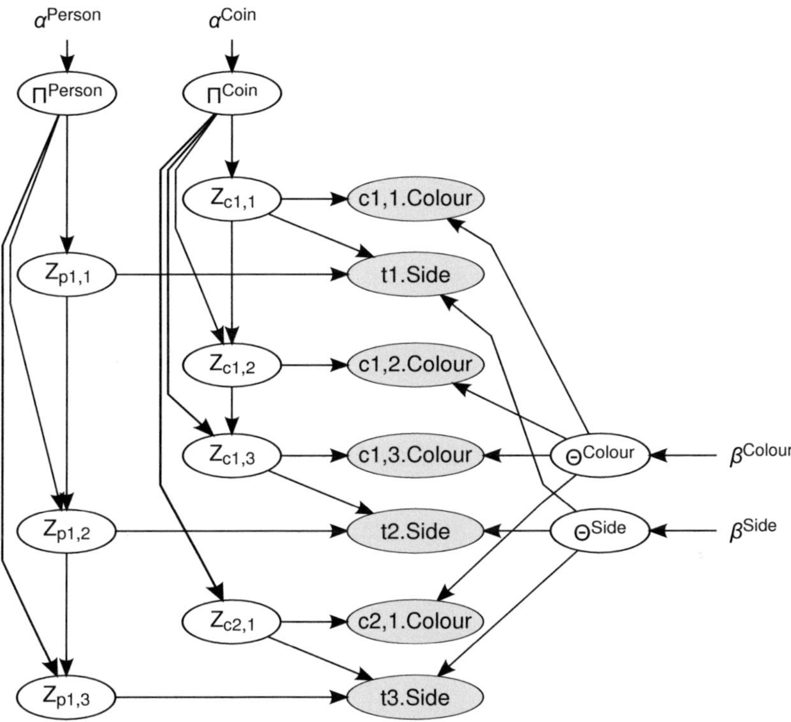

Figure 7.5: Exemplary Bayesian network for the probabilistic model with finite time-varying states. It shows again the hierarchy of random variables and the progress from one state to the next. Compare this diagram with Figure 7.3. This time, the colour of the coin can change over time, for example. c1, c2 and p1 are again coins and a person, t1, t2 and t3 are relationships. The sequence of states is indicated by the number after the comma (e.g., $Z_{c1,1}$, $Z_{c1,2}$ and $Z_{c1,3}$). As in Figure 7.4, I assume Π^{Person} and Π^{Coin} to contain Π'^{Person} and Π'^{Coin}, respectively.

be index with the component indicator and the likelihood of $Z_{e,i+1}$ must be added. The probabilities of $Z_{e,i} = s$ for all $s \in S^R$ then result in

$$P\big(Z_{e,i} = s \mid \mathrm{mb}(Z_{e,i})\big) \propto \pi_{r,s}\,\pi_{s,t} \prod_{Y \in A_e} L^{\mathrm{type}(Y)}\big(a_Y, \theta_{s_Y}^{\mathrm{type}(Y)}\big),$$

where $\pi_{x,y}$ is the element y in the component vector π_x.

The parameter vectors in Π^R can be integrated out as in the time-invariant case. This task starts again with

$$P(Z_{e,i} = s \mid z_{-e,i}, \alpha) = \frac{P(Z_{e,i} = s, z_{-e,i} \mid \alpha)}{P(z_{-e,i} \mid \alpha)}, \tag{7.6}$$

which is free of those parameter vectors. $z_{-e,i}$ is the current joint event of all state variables except for $Z_{e,i}$.

The denominator can directly be obtained by integrating over all parameter vectors π_s. S^R is assumed to contain K states. To ease writing the integral, π denotes the tuple of all parameter vectors π_s and $\mathcal{S}^K$ regards the compound space of all standard simplexes the parameter vectors are defined on. To keep the notation clear, I focus on a single root type R, which should be modelled as proposed in this section, and, thus, omit the superscript R. Then the denominator can be obtained by the following transformations:

$$P(z_{-e,i} \mid \alpha) = \int_{\mathcal{S}^K} P(z_{-e,i} \mid \pi)\, p(\pi \mid \alpha)\, d\pi$$

$$= p_1 \int_{\mathcal{S}^K} \prod_{\substack{e' \in \mathcal{E}^R, j \in \{2,\dots,T_{e'}\} \\ (e',j) \notin \{(e,i),(e,i+1)\}}} P(Z_{e',j} \mid Z_{e',j-1}) \cdot \prod_{x \in S} \prod_{y \in S} \frac{1}{B(\alpha)} \pi_{x,y}^{\alpha_y - 1}\, d\pi$$

$$= p_1 \prod_{x \in S} \int_S \prod_{y \in S} \pi_{x,y}^{n_{x,y}}\, \pi_{x,y}^{\alpha_y - 1}\, d\pi_x$$

$$= p_1 \prod_{x \in S} \frac{B(\alpha + n_x)}{B(\alpha)}.$$

The variables $z_{e',i'}$ for all $e' \in \mathcal{E}^R$ and all $i' \in \{1, \dots, T_{e'}\}$ represent the current value of their corresponding random variables $Z_{e',i'}$ during the sampling. Further $p_1 = \prod_{e' \in \mathcal{E}^R} P(z_{e',1})$ denotes the joint probability of all first variables in the state sequences. It is a constant here. And $n_{x,y}$ is the number of state transitions from the state x to the state y without counting the transitions to and from $Z_{e,i}$. Compare this with the range of the first product in the second line of the above equation. $n_x = (n_{x,y})_{y \in S^R}$ is a vector with all counts that start from a variable in the state x. Altogether the result is the same as in the time-invariant case, except for the additional constant factor p_1 and the conditioning on the previous state.

The numerator in Equation 7.6 can be derived in the same way except that the state transitions to and from $Z_{e,i}$ must be counted. The new counts are represented by the vectors $n_x^{(+Z_{e,i})}$. They match n_x but with the additional transitions $Z_{e,i-1} \to Z_{e,i}$ and $Z_{e,i} \to Z_{e,i+1}$. Numerator and denominator together result in the prior probability of the state $Z_{e,i}$:

$$P(Z_{e,i} = s \mid z_{-e,i}, \alpha) = \prod_{x \in S} \frac{B\left(\alpha + n_x^{(+Z_{e,i})}\right)}{B(\alpha)} \frac{B(\alpha)}{B(\alpha + n_x)} = \prod_{x \in S} \frac{B\left(\alpha + n_x^{(+Z_{e,i})}\right)}{B(\alpha + n_x)}.$$

How this fraction can be reduced, depends on the specific values of the three states r, s and t.

If $r \neq s$, the additional transitions are in two different counting vectors. So the two fractions

$$P(Z_{e,i} = s \mid z_{-e,i}, \alpha) = \frac{B(\alpha + n_r + \hat{e}_s)}{B(\alpha + n_r)} \frac{B(\alpha + n_s + \hat{e}_t)}{B(\alpha + n_s)}$$
$$= \frac{\alpha_s + n_{r,s}}{\alpha_0 + n_{r,0}} \frac{\alpha_t + n_{s,t}}{\alpha_0 + n_{s,0}}$$

remain. Here $\hat{e}_x$ is a standard basis vector with a one at the position of the state $x \in S$. $\alpha_0 = \sum_{y \in S} \alpha_y$ and $n_{x,0} = \sum_{y \in S} n_{x,y}$ for all $x \in S^R$. When comparing this with Equation 7.4 on page 130, the first fraction can be interpreted as the prior probability of the incoming mixture model and the second fraction as the likelihood of the outgoing mixture model.

In the case that $r = s$, the additional transitions affect the same counting vector n_s. Moreover if $s \neq t$, two different counts in the counting vector are changed by one. But the sum of the counts in this vector differs by two. Applying the rule of the gamma function $\Gamma(x + 1) = x\,\Gamma(x)$ recursively, similarly to Equation 7.4 on page 130, leads to

$$P(Z_{e,i} = s \mid z_{-e,i}, \alpha) = \frac{B(\alpha + n_s + \hat{e}_s + \hat{e}_t)}{B(\alpha + n_s)}$$
$$= \frac{(\alpha_s + n_{r,s})\,(\alpha_t + n_{s,t})}{(\alpha_0 + n_{r,0})\,(\alpha_0 + n_{s,0} + 1)}.$$

I chose the indices r and s freely here, because $r = s$. This way of indexing turns out to be useful later on. It also retains the interpretation of the above mentioned first and second fractions.

Finally if $r = s = t$, one counter in one counting vector changes by two. This results in

$$P(Z_{e,i} = s \mid z_{-e,i}, \alpha) = \frac{B(\alpha + n_r + 2\hat{e}_s)}{B(\alpha + n_r)}$$
$$= \frac{(\alpha_s + n_{r,s})\,(\alpha_t + n_{s,t} + 1)}{(\alpha_0 + n_{r,0})\,(\alpha_0 + n_{s,0} + 1)}.$$

Again the indices r, s and t are freely assigned to the variables.

So far, $Z_{e,i}$ had a predecessor and a successor. The cases, if one or both of them are missing, are treated in the following. If $Z_{e,i}$ is the last variable in a state sequence or, in other words, if $i = T_e$, the likelihood of the outgoing transition vanishes. And if $Z_{e,i}$ is the first variable in a state sequence, that is, $i = 1$, a categorical distribution with the parameter vector Π'^R must be taken instead of the mixture distribution with the components Π_s^R. This is the same as in the time-invariant case. The counting vector for the integrated out Pi'^R is regarded as n'^R. If $i = 1 \neq T_e$, the likelihood of the successor must be included in addition to the time-invariant case of Equation 7.4 on page 130.

In summary, six cases must be distinguished depending on, where $Z_{e,i}$ is located within the state sequence and what values its neighbours have. The following equations collect them. For them, I go back to the usual notation with the superscript R to denote the root type under consideration. This notation emphasises the object-relational setting. In addition, I omit factors that are constant for all $s \in S^R$ by using the $\propto$ symbol. If $Z_{e,i}$ is at one end of a state sequence, the probabilities of the categorical distribution are

$$P\left(Z_{e,i} = s \mid z_{-e,i}, \alpha^R\right) \propto \alpha_s^R + n_s'^R \qquad \text{for } i = T_e = 1$$

$$P\left(Z_{e,i} = s \mid z_{-e,i}, \alpha^R\right) \propto \left(\alpha_s^R + n_s'^R\right) \frac{\alpha_t^R + n_{s,t}^R}{\alpha_0^R + n_{s,0}^R} \qquad \text{for } i = 1 \neq T_e$$

$$P\left(Z_{e,i} = s \mid z_{-e,i}, \alpha^R\right) \propto \alpha_s^R + n_{r,s}^R \qquad \text{for } i = T_e \neq 1.$$

In the case that $Z_{e,i}$ is in the middle of a state sequence, the probabilities are

$$P\left(Z_{e,i} = s \mid z_{-e,i}, \alpha^R\right) \propto \begin{cases} \left(\alpha_s^R + n_{r,s}^R\right) \frac{\alpha_t^R + n_{s,t}^R}{\alpha_0^R + n_{s,0}^R} & \text{for } 1 < i < T_e \wedge r \neq s \\[2mm] \left(\alpha_s^R + n_{r,s}^R\right) \frac{\alpha_t^R + n_{s,t}^R}{\alpha_0^R + n_{s,0}^R + 1} & \text{for } 1 < i < T_e \wedge r = s \neq t \\[2mm] \left(\alpha_s^R + n_{r,s}^R\right) \frac{\alpha_t^R + n_{s,t}^R + 1}{\alpha_0^R + n_{s,0}^R + 1} & \text{for } 1 < i < T_e \wedge r = s = t \end{cases}$$

$$\tag{7.7}$$

Because the probabilities depend on s in this case, all three equations are necessary to get one categorical distribution.

Finally the likelihood of the attributes must be included to get the probability of the event $Z_{e,i} = s$ depending on the Markov blanket of $Z_{e,i}$. I do not split this probability up in all the six cases again. Instead I stay with the general formulation

$$P\left(Z_{e,i} = s \mid \mathrm{mb}(Z_{e,i})\right) \propto P\left(Z_{e,i} = s \mid z_{-e,i}, \alpha^R\right) \prod_{Y \in A_e} L^{\mathrm{type}(Y)}\left(a_Y, \theta_{sY}^{\mathrm{type}(Y)}\right).$$

Algorithm 7.3 on the next page shows a possible realisation of the sampling algorithm in a programming style. While the calculations are still simple, many different cases have to be distinguished.

Given the state of the predecessor, the mixture components are independent from one another. How can transfer learning happen then? The

if $i = 1$ **then** $n_s'^R \leftarrow n_s'^R - 1$
else $n_{r,s}^R \leftarrow n_{r,s}^R - 1$

if $i < T_e$ **then** $n_{s,t}^R \leftarrow n_{s,t}^R - 1$

Create p
for $s \in S^R$ **do**
    if $i = 1$ **then** $p_s \leftarrow \alpha_s^R + n_s'^R$
    else $p_s \leftarrow \alpha_s^R + n_{r,s}^R$

    if $i \neq T_e$ **then**
        if $i = 1$ **or** $r \neq s$ **then**
            $p_s \leftarrow p_s \cdot \left(\alpha_t^R + n_{s,t}^R\right) / \left(\text{sum}(\alpha^R) + \text{sum}\left(n_s^R\right)\right)$
        else if $s \neq t$ **then**
            $p_s \leftarrow p_s \cdot \left(\alpha_t^R + n_{s,t}^R\right) / \left(\text{sum}(\alpha^R) + \text{sum}(n_s^R) + 1\right)$
        else
            $p_s \leftarrow p_s \cdot \left(\alpha_t^R + n_{s,t}^R + 1\right) / \left(\text{sum}(\alpha^R) + \text{sum}(n_s^R) + 1\right)$

    for $Y \in A_e$ **do**
        $s_Y \leftarrow$ joint value of Y's parent variables
        $p_s \leftarrow p_s \cdot L^{type(Y)}\left(a_Y, \theta_{s_Y}^{type(Y)}\right)$

$p \leftarrow$ normalise the vector p to sum up to 1
$s \leftarrow$ draw from Categorical(p)

if $i = 1$ **then** $n_s'^R \leftarrow n_s'^R + 1$
else $n_{r,s}^R \leftarrow n_{r,s}^R + 1$

if $i < T_e$ **then** $n_{s,t}^R \leftarrow n_{s,t}^R + 1$

Algorithm 7.3: Sampling of a finite time-varying entity state. This algorithm handles all three cases of a state variable, which can be at the beginning, in the middle or at the end of a state chain. It maintains global count tables n^R and n'^R, instead of counting states like in Algorithm 7.2.

predecessor variables are only fixed temporarily for a Gibbs sampling step, but not for the whole sampling process. Instead every state variable is assigned to every state probabilistically. This way, the evidence associated with each state variable influences all transition probabilities. So transfer learning indeed happens in a limited way. To further smooth the mixture components, α^R could be modelled as a random variable with its own prior distribution. Then α_s^R is a parameter that represents the evidence for all mixture components with regard to the target state, while $n_{r,s}^R$ is a parameter that represents the evidence for a single mixture component. Section 7.3.3 uses this extension for the infinite mixture model.

This subsection features a model, which tracks the inner state of entities. While the observations of an entity were exchangeable in the time-invariant model, the time-varying model of this subsection orders the observations of an entity temporally in form of a Markov chain. This fits well with the concepts of the Enfident Model. The proposed algorithm simplifies the Enfident Model though. The chance of a state transition disregards the time that elapsed between the two associated observations. In contrast to this, Section 6.3 proposes that various more state transitions may have happened when a long time has been elapsed between two observations compared to closely succeeding observations.

The algorithmic realisation integrates out the prior probabilities Π^R. As in the previous section, this mutually connects all state variables that are associated with the same root type. This time however, the preceding and succeeding state variables in the Markov chain are handled differently from all others. They determine that not all state variables of the same root type are connected but only those which have predecessors or successors in the same state.

The whole section covered probabilistic models for the hidden variables, if they have a fixed finite number of states. Two algorithms have been proposed: One assumes that an object's state is time-invariant; the other allows that an object's state changes over time. However the exact number of states is rarely known in practice. How can both algorithms be extended to find the right number of states on their own?

7.3 Realisations with Infinite Mixture Models

This section introduces a technique that assumes a countably infinite number of states while it realises only a finite subset of them. The resulting algorithms do not estimate a suitable number of states. Instead they infer for any number of states at once in a full Bayesian way. There is an implicit probability distribution, whether a state is used or not. The following subsection introduces infinite mixture models and applies them to the time-invariant case. With the concepts of this subsection, the time-varying case can be treated afterwards.

7.3.1 Introduction to the Infinite Mixture Model

In this subsection, the finite mixture model is extended to a model with a countably infinite number of states. In the beginning, the set of states S^R is assumed to have the finite size K. All K elements in the parameter vector α^R should be equal. If α_0^R is the sum of all these values, one element is α_0^R/K. Applying these prerequisites to the prior distribution in Equation 7.4 on page 130 leads to the form

$$P\left(Z_e = s \mid z_{-e}, \alpha_0^R\right) = \frac{\frac{\alpha_0^R}{K} + n_s^R}{\alpha_0^R + n_0^R}.$$

There can be less state variables than possible states. This means, not all states need to be represented in state variables. The states that are currently used are collected in the set $S^{R+} = \{s \in S^R \mid n_s^R \neq 0\}$. As in Section 7.2.2, n_s^R is the number of state variables that have the value s. The other states are in $S^{R-} = S^R \setminus S^{R+}$. Suppose S^{R+} contains J elements. Then the event z_{new} that a state variable Z_e takes on a new state from S^{R-} has the probability

$$P\left(z_{\text{new}} \mid z_{-e}, \alpha_0^R\right) = \sum_{s \in S^{R-}} \frac{\frac{\alpha_0}{K} + n_s^R}{\alpha_0^R + n_0^R} = (K - J)\frac{\frac{\alpha_0^R}{K}}{\alpha_0^R + n_0^R}.$$

In the infinite mixture model, J is kept finite, while $K \to \infty$. Applying this to both of the above equations results in the probabilities

$$P\left(Z_e = s \mid z_{-e}, \alpha_0^R\right) = \begin{cases} \dfrac{n_s^R}{\alpha_0^R + n_0^R} & \text{for an } s \in S^{R+} \text{ and} \\[2ex] \dfrac{\alpha_0^R}{\alpha_0^R + n_0^R} & \text{for a new } s. \end{cases} \tag{7.8}$$

These are the parameters of a state's prior distribution in the infinite mixture model. Neal (1991, 2000) discusses that the limit is well defined. This way of constructing a new state from the already realised states is usually regarded as the *Chinese restaurant process* (CRP) or the *Pólya urn scheme* (see e.g., Pitman, 2006; Xu, 2007; Frigyik et al., 2010).

Note here that the process of sampling a new state variable has a rich-get-richer property. A state that is already used by many other state variables is most likely to be used again. Thus rich states become even richer. This avoids overfitting of the model. With an infinite number of mixture components, data could be explained exactly, if every observation gets its own component. But then, a new observation could not be explained well with the same model. This is an overfitting of the model to the data. The presented sampling schema with the rich-get-richer property suppresses the overfitting by clustering the state variables.

The above equation describes a schema, how a new state is chosen. In a mixture model, every state is associated with a mixture component, that is, with a probability distribution. Consequently a new probability distribution must be instantiated for every new state. If an attribute follows a mixture distribution with components of a parametric family $F(\theta)$, a new distribution can be got by creating a new parameter vector θ_s. This vector is drawn from a prior distribution G_0, that is, $\theta_s \sim G_0$.

Taking all this together, results in a process that provides a countably infinite number of distributions of the same parametric family, all drawn from G_0. The probability of each distribution depends on, how often it is already in use. The algorithmic realisation of this process instantiates a finite number of distributions only. The parameter α_0 of the process controls, how easily a new distribution is generated or, in other words, how sparse the clusters are.

The process could also be said to define a discrete distribution on the continuous space of probability distributions over the observation values. This discrete distribution could be written as

$$G(\theta) = \sum_{k=1}^{\infty} \pi_k \, \delta_{\theta_k}(\theta), \tag{7.9}$$

where every θ regards a probability distribution $F(\theta)$ on the space Ω of the observation values and the θ_k refer to those countably infinite distributions with a probability greater than zero. π_k is the probability of the distribution θ_k. Such distributions over distributions are what the *Dirichlet process* produces. The following expressions sketch the statistical model of the Dirichlet process:

$$G \sim DP(\alpha_0, G_0),$$
$$\phi_i \sim G \quad \text{and}$$
$$y_i \sim F(\phi_i).$$

Both, $\alpha_0 > 0$ and G_0, correspond to the symbols that are already used above. In the context of the Dirichlet process, α_0 is called the concentration parameter and G_0 the base distribution. DP is the distribution of the Dirichlet process: the *Dirichlet measure*. The above expressions simplify the notation a bit for cleanness. G, ϕ_i and y_i are used as random variables and as realisations at the same time. When considering the setting of Section 7.2.2, in which several mixture models are entangled, then y_i is an attribute of the ith mixture model and ϕ_i regards the mixture component of this attribute. In the case of the infinite mixture model, there is a countably infinite number of mixture components identified by $\theta_1, \theta_2, \dots$. ϕ_i can take on any of those values. Expressed with the indicator variable z_i, $\phi_i = \theta_{z_i}$. Thus for indices i and j with $i \neq j$, ϕ_i can be equal ϕ_j, while θ_i and θ_j are always different. The draws from G have a clustering property that is equivalent to that of the Chinese restaurant process.

All this shows that the theory of Dirichlet processes lays out the ground for the infinite mixture model. For completeness, the formal definition of the

Dirichlet process is given in the following, although the above informal introduction is more helpful for infinite mixture models. In the definition, $(\Psi, \mathcal{G})$ is the measurable space of the mixture components. And a finite measurable partition of Ψ is a collection of elements $B_j \in \mathcal{G}$ with $j, k \in J$, if each $B_j \neq \emptyset$, $B_j \cap B_k = \emptyset$, $\bigcup_{j \in J} B_j = \Psi$ and J is a finite set.

Definition 7.3 (Dirichlet process). Let $(\Psi, \mathcal{G})$ be a measurable space, G_0 be a probability measure on this space and α_0 be a positive real number. G is produced by a Dirichlet process with the base distribution G_0 and the concentration parameter α_0, if $(G(B_1), \dots, G(B_k)) \sim$ Dirichlet$(\alpha_0 G_0(B_1), \dots, \alpha_0 G_0(B_k))$ for any finite measurable partition $B_1, \dots, B_k$ (Ferguson, 1973; Van Gael, 2011).

The theory around the Dirichlet process is quite interesting and worth reading, but goes beyond the scope of this dissertation. The article of Ferguson (1973) lays out the basic concepts, while the tutorial of Frigyik et al. (2010) illustrates the various views on the Dirichlet process. The article of Neal (2000) and the dissertation of Xu (2007) describe the Dirichlet process with respect to the infinite mixture model.

This chapter uses a third view on the Dirichlet process: the *stick breaking process* (e.g. Teh et al., 2006; Frigyik et al., 2010). It constructs the probabilities π_k in Equation 7.9. The following equations sketch its statistical model:

$$b_k \mid \alpha_0 \sim \text{Beta}(1, \alpha_0) \quad \text{and}$$

$$\pi_k = b_k \prod_{j=1}^{k-1} (1 - b_j).$$

Beta refers to the beta distribution, which equals to the Dirichlet distribution for the binomial case ($n = 2$ in Definition 7.2).

The idea is to break a stick of length one in two parts. b_1 represents the fraction. The length of the first half is taken as π_1. Then the second half is broken in two parts again. This time, the fraction is b_2 and the length of the first half is assigned to π_2. This procedure is repeated as long as new probabilities are necessary. The remaining second half represents the remaining probability mass. In the mixture model, it corresponds to the probability of the states that are not yet instantiated (the states in S^{R-}).

In summary, the Dirichlet process directly provides a sequence of mixture components for the infinite mixture model. It repeats already used components according to, how often they appeared. The Dirichlet process defines a mathematical concept but no computation schema. Instead other processes are used in algorithms. The stick breaking process describes a computation schema for the probabilities of the mixture components. These probabilities are sometimes necessary. And the Chinese restaurant process adds an intermediate layer in form of the state variables but integrates out the prior probabilities π_k of the mixture components. Its computation schema is easy to use and, thus, the main tool in this section.

7.3.2 Infinite Mixture Models for Time-Invariant Entity Types

The above paragraphs introduced the infinite mixture model. In the following, it is extended to the entity-relationship paradigm in the form proposed by Xu (2007). Several mixture models are entangled so that an observable attribute can depend on the state variables of several entities and that all entities of a certain root type share the same parameters.

Probabilistic model. Let R be a root type with entities, each of which should be modelled with one state variable for all observations, and let $\mathcal{E}^R$ be the set of all entities that are of a type derived from R. These are the entities of interest in this section. The set S^R collects a countably infinite number of states. As in the previous section, I assume it to be the set of natural numbers starting with 1 ($S^R = \{1, 2, ...\}$), just to ease the notation of indices in the following. Then the entities in $\mathcal{E}^R$ are transformed into a probabilistic model in the following way.

Every entity $e \in \mathcal{E}^R$ is associated with a state variable Z_e. It is a random variable that can take on any value in S^R. The state variables follow a Chinese restaurant process (with the distribution named CRP):

$$(Z_e)_{e \in \mathcal{E}^R} \mid \alpha_0^R \sim \text{CRP}(\alpha_0).$$

The basic model here is the same as that of Figure 7.2 on page 125. The hyper-parameter α^R is not a vector though. The Chinese restaurant process needs only the single scalar parameter α_0^R. There are views on the Dirichlet process that make the prior probabilities in Π^R explicit. Then Π^R is an infinite dimensional vector. The Chinese restaurant process works on the collapsed Dirichlet-categorical distribution, which has already been introduced for the finite case as well. The parameter vector Θ^{Colour} of an entity attribute Coin.Colour is an infinite dimensional vector as well. The relationship attribute Throws.Side has two state variables as parents. So its parameter vector Θ^{Side} has $\infty \times \infty$ elements. All these infinite dimensional structures are not explicitly realised during the computations. Rather the algorithm works on a finite subset of them.

If a conjugate prior distribution is used for the parameter vectors Θ^τ, these vectors can be integrated out. This leads to a rather compact model. Indeed this is assumed for the algorithms of the infinite-dimensional case in the following, because the hyper-prior G_0 is explicitly necessary in this case to create new components as already shown in the introduction of the infinite mixture model. The algorithms below are well manageable, when the integration can be performed analytically as with conjugate prior distributions for Θ^τ.

Sampling a state variable. Equation 7.8 on page 143 already gives the prior distribution of Z_e. Naturally one would add the attribute likelihood

$$P(Z_e = s \mid z_{-e}, \alpha_0, A_e) \propto P(Z_e = s \mid z_{-e}, \alpha_0) \prod_{Y \in A_e} L^{\text{type}(Y)}\left(a_Y, \theta_{sY}^{\text{type}(Y)}\right),$$

as it happened in the finite cases, and assume that everything is done. This would require though that all possible $s \in S^R$ are explicitly handled. In the infinite-dimensional case, only the states that are occupied by some state variables should be treated explicitly. They are in the set S^{R+}. Thus the above equation is sufficient for $s \in S^{R+}$. All other states should be treated as a whole. This ensemble of states corresponds to any parameter vector $\theta^{\text{type}(Y)}$ except for those vectors that are associated with the states in S^{R+}.

This means, with $\tau = \text{type}(Y)$, to integrate over the whole space V^τ of the parameter vector θ^τ in the form

$$\int_{V^\tau} L^\tau(a_Y, \theta^\tau)\, G_0^\tau(\theta^\tau)\, d\theta^\tau\ . \tag{7.10}$$

Note that V^τ depends on the parameter vector θ^τ.

Equation 7.10 describes the likelihood that the attribute Y comes from any of the states in S^{R-} by collapsing the attribute distribution F^τ and its prior distribution G_0^τ. If the integral can be solved analytically, the resulting collapsed likelihood depends only on the prior parameters any more. Note that these parameters are not made explicit here. They are part of the function G_0^τ.

The collapsed likelihood can be used for the states in S^{R+} as well. This time, it must include the data that is already associated with a state: It is the a posteriori likelihood. The set $A^\tau_{-Y|s_Y}$ denotes this data. It contains all attributes of type τ that are associated with the same component s_Y except for the attribute Y. The a posteriori likelihood corresponds to the a posteriori probability

$$P(Y \mid A_{-Y|s_Y}, G_0) = \frac{P(Y, A_{-Y|s_Y} \mid G_0)}{P(A_{-Y|s_Y} \mid G_0)}\ .$$

Transforming this to a likelihood, results in the function

$$K_{s_Y}^\tau(Y) = \frac{\displaystyle\int_{V^\tau} L^\tau(a_Y, \theta^\tau) \left(\prod_{Y' \in A^\tau_{-Y|s_Y}} L^\tau(a_{Y'}, \theta^\tau) \right) G_0^\tau(\theta^\tau)\, d\theta^\tau}{\displaystyle\int_{V^\tau} \left(\prod_{Y' \in A^\tau_{-Y|s_Y}} L^\tau(a_{Y'}, \theta^\tau) \right) G_0^\tau(\theta^\tau)\, d\theta^\tau}\ .$$

It is the a posteriori likelihood of the attribute Y of type τ given the mixture component s_Y. Or in other words, it is the likelihood of the attribute Y for the mixture component s_Y given all data that is also associated with this component and expressed in terms of the prior parameters instead of θ^τ.

If G_0^T is the conjugate prior of the distribution, to which the likelihood L^T belongs, $K_{s_Y}^T$ can be obtained analytically. Note that $A^T_{-Y|s_Y}$ is empty in the case of a new state $s \in S^{R-}$ for Z_e. With this empty set, $K_{s_Y}^T$ reduces to the integral of Equation 7.10, because the products are empty and the integral over G_0^T in the denominator results in 1. Thus $K_{s_Y}^T$ fits for the cases that s is from S^{R+} and from S^{R-}.

With this likelihood function, the a posteriori distribution of Z_e can be expressed compactly. It has the probabilities

$$P(Z_e = s \mid z_{-e}, \alpha_0^R, \mathcal{A}_e) \propto \begin{cases} n_s^R \prod\limits_{Y \in A_e} K_{s_Y}^{\mathrm{type}(Y)}(Y) & \text{for an } s \in S^{R+} \text{ and} \\ \alpha_0^R \prod\limits_{Y \in A_e} K_{s_Y}^{\mathrm{type}(Y)}(Y) & \text{for a new } s. \end{cases}$$

The set A_e contains all attributes Y that are associated with the entity e either directly or through a relationship ($A_e = \{Y \mid e \in \text{entities}(Y)\}$). And $\mathcal{A}_e$ denotes a set, which contains all the attributes in A_e plus all attributes that are of the same type as those in A_e. This way, $\mathcal{A}_e$ also contains all attributes of the sets $A^{\mathrm{type}(Y)}_{-Y,s_Y}$.

Algorithm 7.4 on the next page expresses this sampling schema in a programming style. Altogether the algorithm is similar to the finite case. Only few additional operations are necessary to include the probability of all the states that are not instantiated in a state variable. However during the sampling, the algorithm randomly creates many states that are used by one or two state variables only. So the total number of instantiated states might be higher compared to a finite mixture model with a well estimated number of states. On the other hand, the estimation of the number of states is unnecessary here.

7.3.3 Infinite Mixture Models for Time-Varying Entity Types

The finite time-varying case in Section 7.2.3 uses a Markov chain in a similar way as a hidden Markov model. Therefore I roughly introduce the infinite hidden Markov model first and extend it then for the object-relational domain.

$n_s^R \leftarrow n_s^R - 1$

if $n_s^R = 0$ **then** remove unnecessary elements

// Probabilities for states that are already in use

$p \leftarrow n^R$

for $s \in S^{R+}$ **do**

    for $Y \in A_e$ **do**

        $s_Y \leftarrow$ joint component indicator with $Z_e = s$

        $p_s \leftarrow p_s \cdot K_{s_Y}^{\text{type}(Y)}(Y)$

// Overall probability for all unused states

$q \leftarrow \alpha_0^R$

for $Y \in A_e$ **do**

    $s_Y \leftarrow$ joint component indicator for a new state in Z_e

    $q \leftarrow q \cdot K_{s_Y}^{\text{type}(Y)}(Y)$

Append value q to the vector p

$p \leftarrow$ normalise the vector p to sum up to 1

$s \leftarrow$ draw from Categorical(p)

if $s \in S^{R+}$ **then**

    $n_s^R \leftarrow n_s^R + 1$

else

    $s \leftarrow$ a new appropriate state value *// possibly reuse old values*

    Extend n^R to contain $n_s^R = 1$

Algorithm 7.4: Sampling of an infinite time-invariant state variable. s is the variable for the current state of the random variable Z_e and n^R is a one-dimensional list of the state counts for all states in S^{R+}.

Idea of an infinite hidden Markov model. The section about the finite time-varying case already mentions that the hyper parameters α^R can be understood as random variables in order to learn across mixture components. This is necessary here. To do so, the parameter vector α^R is split in two parts: $\alpha^R = \alpha_0^R \beta^R$. β^R is a unit vector and α_0^R a scaling factor. Then β^R is turned into a random variable denoted as B^R and gets a Dirichlet prior distribution with parameters, that all have the value γ^R/K. K is as always the number of states in S^R and γ^R is another scaling factor. Both, α_0^R and γ^R can be random variables with a gamma distribution as, for example, Teh et al. (2006) show. For this chapter, it is sufficient to model them as constants. The extended finite-dimensional model can be summarised with the expressions

$$B^R \sim \text{Dirichlet}\big((\gamma^R/K, \dots, \gamma^R/K)\big),$$
$$\Pi_s^R \mid B^R = \beta^R, \alpha_0^R \sim \text{Dirichlet}\big(\alpha_0^R \beta^R\big) \text{ and}$$
$$Z_{e,i} \mid Z_{e,i-1} = r, \Pi^R = (\pi_s)_{s \in S^R} \sim \text{Categorical}(\pi_r)$$

with $s, r \in S^R$.

What does this form of coupling bring about? Equation 7.7 on page 139 contains terms of the form $\alpha_s^R + n_{r,s}^R$. Through the evidence variable $n_{r,s}^R$, every component of the transition distribution learns on its own. It only uses the evidence of the given predecessor r. In contrast, α_s^R is a parameter solely depending on the current state. Making this parameter a random variable learns a probabilistic effect for the state s, which is common to all the predecessors. Thus when using the extension above, $\alpha_s^R + n_{r,s}^R$ combines the learning of source- and target-related correlations for the transitions in the Markov chain.

The new model connects two Dirichlet distributions. It is a hierarchical Dirichlet model. With $K \to \infty$, it results in a hierarchical Dirichlet process, which has been proposed by Teh et al. (2006). Van Gael et al. (2008) further detailed its application on the infinite hidden Markov model. The idea is the following. Given $Z_{e,i-1}$, the indicator variable $Z_{e,i}$ together with the attribute variable Y_i makes up a mixture model. The state transition probabilities $\pi_{Z_{e,i-1},s}$ for $s \in S^R$ are the mixture weights. In the infinite case, the mixture model can be represented by a Dirichlet process. Every state in $Z_{e,i-1}$ regards its own mixture model and thus its own Dirichlet process. So there is an infinite set of Dirichlet processes or, in other words, an infinite set of mixture models for an attribute. With a hierarchical Dirichlet process, these attribute-related processes are coupled in the way that they share the same mixture components but with different mixture weights. The mixture components that they share come from another Dirichlet process. This parent process also produces an infinite-dimensional vector B^R of probabilities, around which the mixture weights of the child processes are centred. B^R corresponds to the finite-dimensional vector B^R in the equations above.

The theory about hierarchical Dirichlet processes is so complex that presenting it here would loose the scope of this work. Therefore I point the reader to the mentioned papers. In the following, I introduce an object-relational model based on these concepts. I present it without the theory of hierarchical Dirichlet processes. Instead I build on top of the previous sections to point out the idea.

Probabilistic model.   Let R be a root type whose entities should be modelled with a time-varying state. $\mathcal{E}^R$ denotes the set of all entities that are of a type derived from R. These are the entities of interest in this section. S^R is a countably infinite set of all states. Then the entities in $\mathcal{E}^R$ are transformed into a probabilistic model in the following way.

For every observation $i = 1, 2, \ldots, T_e$ of every entity $e \in \mathcal{E}^R$, add a random variable $Z_{e,i}$ that represents the inner state of the entity during the observation. The state variables $Z_{e,i}$ of the entity e for all i make up a Markov chain. Thus $Z_{e,i}$ with $i \neq 1$ depends only on its predecessor $Z_{e,i-1}$. The collapsed Dirichlet-categorical distribution for the infinite hidden Markov model has no name in the literature. In the following, I call it the time-varying Chinese restaurant franchise (TVCRF). It is an extension of the Chinese restaurant process and generalises Equation 7.7 to the infinite-dimensional case. TVCRF' is the corresponding process for the first state variable in a state sequence. These processes are coupled with a random variable B^R. B^R is an infinite-dimensional probability vector, which is associated with the root type R of the entity. It is generated by a stick breaking process with the distribution named Stick (Section 7.3.1). The time-varying Chinese restaurant franchise and the stick breaking process have the root-type-specific parameters α_0^R and γ^R, respectively. Both are positive real numbers. The following expressions summaries the random variables and parameters:

$$B^R \sim \text{Stick}\left(\gamma^R\right),$$

$$Z_{e,1} \mid z_{-e,1}, B^R = \beta, \alpha_0^R \sim \text{TVCRF}'\left(\alpha_0^R, \beta\right) \quad \text{and}$$

$$Z_{e,i} \mid z_{-e,i}, B^R = \beta, \alpha_0^R \sim \text{TVCRF}\left(\alpha_0^R, \beta\right) \quad \text{for } i \neq 1.$$

Here $z_{e,i}$ is the current joint event of all state variables of the entities in $\mathcal{E}^R$ except for $Z_{e,i}$.

This probabilistic model is comparable with the finite-dimensional model of Section 7.2.3 but extends it in the following ways. The model uses a countably infinite number of states, but instantiates only a finite subset of it. Consequently the transition matrix and the set of attribute mixture components are infinite dimensional as well. An additional random variable B couples the components of the transition distribution. It is an infinite-dimensional vector of prior probabilities.

What features does the model offer in the end? With the hidden state variables, it learns an individual mixture model for every observation of an entity. The state transition probabilities and the mixture parameters for the attributes are shared variables for all entities of a root type. This way, the model transfers evidence between entities. For a trust model, this means that the model can learn the behaviour of, for example, every cooperation partner individually. But it shares behavioural patterns across all cooperation partners. Through the additional hierarchy level with the variable B, evidence is shared between the transition components. This can be realised in a similar way for the mixture distributions of the attributes. All these are important features for a trust model in order that it can support the ability to decide under high uncertainty. On the downside, the state transition probability only depends on the root type. In some scenarios, transition probabilities could be desirable that are individual for each entity. Partly the proposed model can compensate this missing feature in the way that a state variable is in all states, just with different weights. This way, the state trajectory of different entities can be individual. If this is not sufficient, another Dirichlet process level can be inserted to group all observations of an entity.

So far, I introduced the setting. The remaining section shows shortly how to sample the state variables and the hyper-parameter B. For details, please refer to the papers mentioned above.

Sampling a state variable　The distribution over the first state variable in a sequence (Π'^R) can be understood as just another mixture component in the transition matrix (Π^R). The component is selected for all first state variables with probability one and for all other state variables with probability zero. This can be realised in the algorithm by adding a constant $Z_{e',0}$ to all entities e' with a state value, which is not in S^R. For example, if $S^R = \{1, 2, ...\}$, then $Z_{e,0}$ can be set to zero. This simplifies the algorithm, because fewer branches are necessary. Consequently only Equation 7.7 on page 139 is the relevant part from Section 7.2.3 regarding the prior distribution.

In the following, the state variable $Z_{e,i}$ for $i = 1, ..., T_1 - 1$ should be sampled. The variables r, s and t should be the values of $Z_{e,i-1}$, $Z_{e,i}$ and $Z_{e,i+1}$, respectively. Let $z_{-e,i}$ be the joint event of all other state variables, that is, all $Z_{e',j}$ with $e' \in \mathcal{E}^R, j \in \{0, ..., T_{e'}\}$ and $(e', j) \neq (e, i)$. Similarly to the infinite-dimensional time-invariant algorithm, there is a set S^{R+} for all states that are used in $z_{-e,i}$ and a set S^{R-} for all other states. These other states are only treated as a whole. Moreover let β^R be the current realisation of the random variable B. Then the prior probabilities of Equation 7.7 can be reformulated to

$$
P\left(Z_{e,i} = s \mid z_{-e,i}, \alpha_0^R, \beta^R\right) \propto
\begin{cases}
\left(\alpha_0^R \beta_s^R + n_{r,s}^R\right) \dfrac{\alpha_0^R \beta_t^R + n_{s,t}^R}{\alpha_0^R + n_{s,0}^R} & \text{for } r \neq s \in S^{R+} \\[2mm]
\left(\alpha_0^R \beta_s^R + n_{r,s}^R\right) \dfrac{\alpha_0^R \beta_t^R + n_{s,t}^R}{\alpha_0^R + n_{s,0}^R + 1} & \text{for } r = s \neq t \\[2mm]
\left(\alpha_0^R \beta_s^R + n_{r,s}^R\right) \dfrac{\alpha_0^R \beta_t^R + n_{s,t}^R + 1}{\alpha_0^R + n_{s,0}^R + 1} & \text{for } r = s = t \\[2mm]
\alpha_0^R \beta_u^R \beta_t^R & \text{for a new } s.
\end{cases}
$$

In the last case, $\alpha_0^R \beta_u^R$ is the incoming probability to get from r in any of the states in S^{R-} and β_t^R is the outgoing probability to reach the successive state t. For the last state variable Z_{e,T_e} in a chain, the second factor of the first three cases is omitted, because there is no outgoing transition. In the consequence, the above probability has only two cases for Z_{e,T_e} any more.

The probabilities of the sampling distribution

$$P(Z_{e,i} = s \mid mb(Z_{e,i})) \propto P\left(Z_{e,i} = s \mid z_{-e,i}, \alpha_0^R, \beta^R\right) \prod_{Y \in A_{e,i}} K_{sY}^{type(Y)}(Y)$$

just contain the likelihood of the attributes in addition. $K_{sY}^{type(Y)}$ is defined as in the previous section. Algorithm 7.5 on the following page realises this schema in programming style. It is straightforward except for the case when a new state is created. How is this done, since the number of states for all state variables and for the parameter vector B must match?

A stick breaking process generates the variable B with the beta distribution as described in Section 7.3.1. β_u^R represents the size of the remaining stick or the remaining probability mass. If a new state is sampled, β_u^R is broken by the random factor $b \sim Beta(1, \gamma)$ in the way that

$$\beta_{s_{new}}^R = b\,\beta_u^R \qquad \text{and}$$
$$\beta_u^R = \beta_u^R - \beta_{s_{new}}^R.$$

This schema ensures that $\beta^R = \left(\beta_1^R, \dots, \beta_K^R, \beta_u^R\right)$ is Dirichlet-distributed. In the case here, this the whole stick breaking computation, because no evidence is yet available for β_u^R, since it represents the unrealised states. Algorithm 7.5 contains this update of β^R as well. It also notes that the count table must be updated if a new state is instantiated.

In this hierarchical setting, the parameter vector B^R must be sampled as well. It cannot be integrated out analytically. The following paragraphs reproduce the sampling schema from Teh et al., 2006.

Sampling B^R. Before sampling B^R, the data structures should be cleaned up. All states that are not used any more ($n_{r,s}^R = 0$) should be removed.

To get the probability distribution of B^R, an auxiliary random matrix M^R is necessary. In Teh et al.'s (2006) Chinese restaurant franchise schema, M_{jk}^R

$n_{r,s}^R \leftarrow n_{r,s}^R - 1$
if $i \neq T_e$ **then** $n_{s,t}^R \leftarrow n_{s,t}^R - 1$

Create p
for $s \in S^{R+}$ **do**
$\quad p_s \leftarrow p_s \cdot \left(\alpha_0^R \beta_s^R + n_{r,s}^R \right)$

$\quad$ **if** $i \neq T_e$ **then**
$\quad\quad$ **if** $r \neq s$ **then**
$\quad\quad\quad p_s \leftarrow p_s \cdot \left(\alpha_0^R \beta_t^R + n_{s,t}^R \right) / \left(\alpha_0^R + \mathrm{sum}\left(n_s^R \right) \right)$
$\quad\quad$ **else if** $s \neq t$ **then**
$\quad\quad\quad p_s \leftarrow p_s \cdot \left(\alpha_0^R \beta_t^R + n_{s,t}^R \right) / \left(\alpha_0^R + \mathrm{sum}\left(n_s^R \right) + 1 \right)$
$\quad\quad$ **else**
$\quad\quad\quad p_s \leftarrow p_s \cdot \left(\alpha_0^R \beta_t^R + n_{s,t}^R + 1 \right) / \left(\alpha_0^R + \mathrm{sum}\left(n_s^R \right) + 1 \right)$

$\quad$ **for** $Y \in A_e$ **do** $p_s \leftarrow p_s \cdot K_{s_Y}^{\mathrm{type}(Y)}(Y)$

$q \leftarrow \alpha_0^R \beta_u^R$
if $i \neq T_e$ **then** $q \leftarrow q \cdot \beta_t^R$

for $Y \in A_{e,i}$ **do**
$\quad s_Y \leftarrow$ joint component indicator for a new state in $Z_{e,i}$
$\quad q \leftarrow q \cdot K_{s_Y}^{\mathrm{type}(Y)}(Y)$
Append q to p

$p \leftarrow$ normalise the vector p to sum up to 1
$s \leftarrow$ draw from Categorical(p)

if $s \notin S^{R+}$ **then**
$\quad$ Extend n^R to contain $n_{x,s}^R = 0$ and $n_{s,x}^R = 0$ for all $x \in S^{R+}$
$\quad b \leftarrow$ draw from Beta$\left(1, \gamma^R \right)$
$\quad \beta_s^R \leftarrow \beta_u^R b$
$\quad \beta_u^R \leftarrow \beta_u^R - \beta_s^R$
$\quad$ Add s to S^{R+}

$n_{r,s}^R \leftarrow n_{r,s}^R + 1$
if $i \neq T_e$ **then** $n_{s,t}^R \leftarrow n_{s,t}^R + 1$

Algorithm 7.5: Sampling the state variable in the infinite time-varying case. This algorithm combines the techniques from Algorithms 7.3 and 7.4.

represents the number of tables in the restaurant j with the dish k. The dish k is the state of $Z_{e,i}$, while the restaurant j is the state of $Z_{e,i-1}$. M^R has a categorical distribution with the probabilities

$$P\left(M_{jk}^R = m \mid z, m_{-jk}, \beta^R\right) = \frac{\Gamma\left(\alpha_0^R \beta_k^R\right)}{\Gamma\left(\alpha_0^R \beta_k^R + n_{j,k}^R\right)} \; s\left(n_{j,k}^R, m\right) \left(\alpha_0^R \beta_k^R\right)^m.$$

The symbol z denotes the joint event of all state variables in $\mathcal{E}^R$ and m_{-jk} regards the joint event of all elements in M^R except of M_{jk}^R. $s(n, m)$ are unsigned Stirling numbers of the first kind.

 With this random matrix, the probability distribution of B^R is simply an updated Dirichlet distribution:

$$\left(\beta_1^R, \dots, \beta_K^R, \beta_u^R \mid M^R, \gamma^R\right) \sim \text{Dirichlet}\left(m_1, \dots, m_K, \gamma^R\right),$$

where m_i is the sum over the column i in the current value of M^R ($m_i = \sum_{s=1}^{K} m_{si}$).

 I omit the programming code for the sampling of both variables. It would not add much to the equations above. In summary, the hierarchical setting of the model in this section requires to sample two additional variables. Furthermore the algorithm for sampling a state variable is more complicated. But still it requires just one loop over S^{R+} and a sub-loop over the attributes. Only in case, that a new state is created, all data structures must be extended.

7.4 Summary

At the beginning of this dissertation, trust has been introduced as a mechanism that learns model of others' behaviour. What does this mean for an algorithm? When the trusting agent knows little about the world, few other agents and few trust experiences, it has a simple model only. With increasing number of experiences and interaction partners, the model should become more extended, detailed and complex. Learning a model of others' behaviour also means that a trust algorithm should be able to flexibly learn any behavioural pattern. It should make as little assumptions about the others' behaviour as possible. At the same time, it should learn every behavioural pattern as exact as possible.

So a trust algorithm should be able to handle complex and simple scenarios, that is, little and much evidence. And it should be unspecific regarding the possible behaviour pattern, but learn every specific behavioural pattern exactly. The poles of both statements seem incompatible for algorithms. For example, if an agent models a discrete attribute with six levels, it can apply a categorical distribution with these six levels to it. This is the most generic model schema, as it provides five degrees of freedom for the probabilities. Now the agent has three pieces of evidence. It can learn the categorical distribution with this evidence, but the result will not be reliable. It is likely to change with more evidence, because three random points are too few for five degrees of freedom. If the agent would know that the discrete values come from the discretion of a Gaussian distributed continuous random magnitude, the probabilistic model would have two degrees of freedom only. Thus it can find the right distribution with little evidence more easily. In addition, the degrees of freedom add up over all mixture components. In general, a complex probabilistic model has many ways to explain few data. For a sampling algorithm, the consequence is that the sampling space is large and, thus, the algorithm could fail to converge. Similar problems may occur with maximum likelihood or maximum a posteriori learning. Altogether a trust algorithm must be flexible and specific at the same time. This is an inherent property of the trust mechanism.

The proposed algorithms include several means to realise this property. Through their entity-relationship paradigm they can scale from simple to complex scenarios: With an increasing number of entities, the structure of the underlying probabilistic model increases as well. And the infinite-dimensional algorithms adjust the number of states as needed; they are non-parametric. This is also a kind of structure learning, as the number of states matches with the number of distributions in the mixtures. With an increasing amount of evidence, they can distinguish more and more mixture components. While the probabilistic models of the attributes are important for this property of the trust mechanism as well, they have been omitted, because they are application specific and not at the core of the trust model.

Further advantages of the proposed algorithms are already mentioned throughout this chapter several times. Especially the algorithms try to transfer the available evidence between all entities. This is important to judge under high uncertainty – another inherent requirement of the trust mechanism.

Besides these advantages, the proposed algorithms were mainly selected for this dissertation, because they suit well to explain many algorithmic facets of trust and the Enfident Model. They are based on generative probabilistic models, which reflect assumptions about trust in a similar way as the Enfident Model and can well be visualised. This meets the scope of this dissertation to improve the understanding and modelling of trust between cooperating technical systems.

However explaining the world is also one of the drawbacks of generative models. The main purpose of an inference algorithm is to match the data and not any assumption about the world. For this reason, it would be interesting for the future to apply discriminative regression and clustering techniques on the Enfident Model. This could also lead to a higher computational speed, since especially the sampling algorithms for the infinite-dimensional mixtures are very slow. In addition to these drawbacks, the proposed time-varying algorithms do not include the progressing time in the transition probabilities of the Markov chain. This differs from the Enfident Model as explained in Section 6.3. Moreover the algorithms just compute probabilistic trust. They disregard potential gain and loss. They could be extended for interest-related trust though. Finally Gibbs sampling can fail to converge to the right distribution. This can also be a problem in real applications. Despite all these problems, the proposed algorithms still have good practical advantages and fit well with the aim of this dissertation.

8 Evaluation Method

To show trust in action, I apply it to communicating vehicles. The following three sections introduce the scenario more in detail. Why did I select this scenario for the evaluation of the Enfident Model? On the one hand, applying trust here is reasonable: Cognitive vehicles have the necessary ability to judge cooperation outcomes; and they benefit from a trust model as the next section explains. On the other hand, the scenario can be simulated well: The social structure can be constructed as shown in Section 8.4.1; it determines who is interacting with whom. And the message exchange and information processing can sufficiently be reproduced in a simulation too.

While an application scenario is helpful for the concreteness and vividness of the evaluation, what should be the scope of an evaluation in this dissertation? The aim of this dissertation is not to find the fastest trust algorithm, for example. Rather it contributes to the understanding and modelling of trust between technical systems on a conceptual level. Thus the evaluation should show that the postulated requirements and the Enfident Model lead to intuitive and consistent results. Moreover the Enfident Model and the realising algorithms should fulfil the requirements.

Consequently the requirements are connected with application-specific use cases in Section 8.2. These use cases are in turn transformed in simulation scenarios as described in Sections 8.5.1 and 8.5. The simulation scenarios are comparable to tests. Simulating them leads to test results presented in Chapter 9. That chapter also relates the results back to the requirements and verifies them regarding practical expectations.

8.1 The Inter-Vehicular Communication Scenario

Consider a scenario in which cars continuously talk with each other while driving around. These future cars can perceive, reason, learn, plan and act in a way that they understand the surrounding traffic scene. In the literature, they are sometimes called *cooperative cognitive automobiles* (Stiller et al., 2007). Compare this with cognitive technical systems in general as introduced in Chapter 1.2. Such vehicles are expected to improve the traffic efficiency and safety. While they may still be driven by a human, they can widely assist the driver. To extend their perception range and their capabilities, they exchange all kinds of model information through a vehicular network. *In this setting, a trustworthy cooperation partner is a vehicle that sends accurate information.* Note that vehicles also cooperate in the way that they slow down or change lanes because of other vehicles on the road. This form of cooperation is not considered here.

The following listing collects examples of information that the vehicles store in their knowledge base and exchange through the network (Enkelmann, 2003; Matheus et al., 2004; CAR 2 CAR Communication Consortium, 2007; Kranz, 2008). Its classification is arbitrary. It should just make aware of the various kinds of applications associated with a vehicular network.

Long-term information. Structural alterations, permanent changes in the traffic regulations (e.g. traffic signs), new or removed points of interest (like petrol stations, hotels or museums).

Medium-term information. Temporary changes in the traffic regulation (e.g. traffic signs because of special events), hazardous locations (e.g. dirt or oil on the road), content changes of a point of interest (e.g. prices).

Short-term information. Hazardous locations (e.g. fog, aquaplaning), obstacles, free parking space, level of service, traffic congestions.

Situational information. Green light optimal speed advice, location and velocity of traffic participants (for collision warning, collision preparation and merging assistance to join flowing traffic), location and route of emergency vehicles.

In the considered scenario, a vehicle has two sources of information: It continuously scans its environment for interesting features (named *observations*) and it receives information about the environment from other vehicles (named *reports*). All reports must be evaluated and some decisions must be derived from them, like: Should I take the information as correct and integrate it into my knowledge base? Should I recommend the driver to choose a different route, because the received information suggests that? Should I warn the driver to slow down, because the other vehicle informed me about bad road conditions ahead?

While vehicular networks are already there in industrial projects, they will become even more important for vehicles with cognitive capabilities. This is because such a vehicle will be able to gather more information and to understand a situation more thoroughly – it is better able to cooperate. With the obtained information, it advises the driver and takes control in emergency situations. But can the vehicle be sure, that the information received from other vehicles is correct?

Using a trust mechanism offers a cooperating vehicle various advantages for the communication management (following Chapter 1.2): The vehicle can better judge the reliability of received information; as a consequence, it can more certainly know whether available evidence is sufficient for a decision; and it can more reliably select a cooperation partner to ask for further information about a subject. In the consequence of this improved communication management, the information exchange might become more reliable and efficient, and the inner models of the vehicles about their environment might get more accurate and complete.

Note that the scenario in this dissertation is simplified. A vehicular network involves not only vehicles but also fixed roadside-units. They are, for example, associated with a traffic light. Vehicles may also exchange information with the Internet and with radio stations. In addition, the considered setting is completely decentralised. A real vehicular network will probably combine distributed and central elements. Finally I also disregard privacy aspects. All this would induce additional complexity in the application scenario without any benefit for the evaluation. Therefore I chose the simplified scenario here.

8.2 Use Cases for the Trust Model

This section derives use cases for the scenario described above. It presents few use cases only. They should illustrate the main ideas about trust in this scenario, cover all the requirements of Chapter 4 and be reproducible in the simulation.

The common theme of the use cases is that somehow wrong information got into the network and the trust model should support the knowledge management to maintain a correct knowledge base. Consequently the use cases are sorted according to the reason for the wrong information. In addition, the listing below notes the desired behaviour of the trusting vehicle. In this way, the text can directly link to the requirements.

Usual behaviour use cases.   Even if every car works as intended – no defects, no manipulations –, the trust mechanism must capture the behaviour of other cars, because technical systems are always limited in their abilities.

1. Cars have differently restricted sensory capabilities. Therefore they sometimes miss to report an event (like a new sign) or spread out wrong information (e.g. a wrong level of service). The error rate depends on the car and on the involved sensors.

   The trusting car must know how correct the values of the other car usually are. This property refers to the ability or competence of the other car following the nomenclature of Falcone et al. (2003) and of Chapter 6. The ability depends on the task to perform (for example, on the quality of the speed sensor or the pedestrian detection system). Thus the trusting car should develop trust depending on the cooperation partner and the cooperation task (Requirement 1). This also includes situational properties like the fact that the detection of a traffic sign was at night (darkness) or during the day (daylight).

   Typically the task-specific ability cannot be known in advance (e.g. depending on the vehicle model). Some properties of the cooperation partner may be a good indicator in the beginning, but the individual, task-specific ability can only be learned over several interactions (Requirement 4).

2. A car enters the vehicular network and starts to participate regularly. Prior to that, it was not equipped with a vehicular network adapter.

   The trusting car should learn, how new participants perform typically by generalising over all new participants, with which it has interacted before (generalisation as demanded in Requirement 6). It can then adjust its trust in this participant in the course of further interactions to a value, which is specific for that cooperation partner (specialisation as demanded in Requirement 6 and experience-based learning as demanded in Requirement 4).

3. At a crowded place, a car receives many messages of various kinds. Some are safety-relevant like the velocity vectors of other vehicles; some are of low importance like offers of a new hotel nearby. Slowly the car's input queue is filling up, because it fails to process the messages as quickly as new ones come in. So it decides to drop all messages of low importance.

   This use case describes the actions of the trusting car. The car drops messages, thus it refuses cooperation. It does so to handle its difficult situation best. The correctness of the dropped messages or the trustworthiness of their senders does not play any role. The decision what to do with the messages can be made without trust considerations (Requirement 3). This use case shows the possible divergence between an inner trust state and a trust-related decision.

4. A car wants to know, how good the traffic is on the highway it is going to. It meets a moped and some other cars that come from this direction. It has not met all of them before.

   In this use case, the trusting car uses trust to decide whether to start cooperation with the moped. Because a moped must not drive on a highway, the car does not need to ask it for the desired information. Instead the trusting car should contact some of the other cars. Thus trust arises just from logical considerations without prior experiences (Requirement 4).

Use cases with defects.   In the use cases above, the performance of the cooperation partner is assumed to be system inherent and thus constant. If a defect happens, the performance varies in time. The variations could happen slowly or suddenly. A trust mechanism should provide means to model the time variance. The following list gives two examples.

5. A car has a defect speed sensor. It displays a deviation of 15 km/h. Therefore it wrongly reports levels of service and traffic congestions. After a while, the defect is detected and repaired.

   The trusting car should decrease the trust in the partner, the more it is sure about its disability. This decrease should depend on the kind of information though. Some kinds rely on the defect sensor, others do not. When later, the good performance is back, it should rehabilitate the old trust level (Requirement 7).

6. Again a car has a defect speed sensor. This time, it detects the defect automatically after one day by comparing the own speed with the values reported by neighbouring cars. As a consequence, it tells the others that its speed reports were wrong since one day. When the defect is repaired it also mentions that.

   The trusting car should re-evaluate all reasoning operations that were affected by the wrong speed values. It should also reconsider its trust and respect the honesty of the cooperation partner. Trust should always be reconstructed from stored evidence every time it is needed. This is necessary to handle changes in the evaluation of former evidence as described in this use case. Pearl (1988) emphasises that it can always be necessary in reasoning to retract a conclusion when new evidence is available – reasoning is non-monotonic (Requirement 5).

Use cases with malicious behaviour.   A malicious node in the vehicular network can be a vehicle that has been manipulated somehow or a faked vehicle like a simple notebook with a cracked access to the network. The motivations for such attacks on the vehicular network can be quite different: An operator of a point of interest could spread outdated information about competitors with a manipulated roadside-unit. Others could try to make up a virus-based infrastructure, like that for spam in the Internet, in order to present context-related advertisement to the driver. These two examples show that there is some economic interest in inducing unwanted information into a vehicular network. Besides this economic motivation, a hacker may also act for several non-commercial reasons like showing his skills. Raya (2009, pp. 10–13) gives a systematic overview. She classifies the attacker according to his network membership, motivation, method and scope. Based on this structure, she introduces various kinds of attacks. I do not recall them here. The use cases in the following should just illustrate some trust-related problems without being comprehensive.

There is a schema that distinguishes the use cases introduced so far from those in the following. The above use cases discuss the reliability of the co-operation partner regarding its abilities, considering the sensory systems. In contrast, the manipulations below regard the control system of the co-operation partner. So the abilities are still the same, but the willingness to cooperate changed (see Requirement 1 and Chapter 6).

7. Every Wednesday, a car reports that a certain car park is full to ensure it can park there. In addition, it reports bad recommendations of other cars that mention free parking space there. Besides this malicious behaviour, the car acts trustworthy.

   The trusting car should consider situational attributes that could correlate with the other's willingness. For example, these attributes could be the time of interaction, the expected parking space or the amount of bad recommendations (Requirement 1).

   Even with the consideration of willingness-related attributes, the behaviour may still seem random. As a consequence, the trusting car should also evaluate, how well it can predict the other's behaviour. Cooperation can only be reliable if the trust mechanism can predict the cooperation outcome (Requirement 2).

   Finally this use case also indicates that recommendations may be wrong. Scenario 9 covers this issue.

8. An attacker designs a software virus that infects cars to spread advertisements with a high trustworthiness. A certain vehicle gets infected. Half a year later, the virus is removed during a regular inspection.

   In this use case, the remarkable pattern might be that the infected car sends many messages with a message type (point of interest) that is not important compared to other more traffic related message types. The infected car misuses the infrastructure and the other cars' knowledge base.

   As a consequence, the trusting car should not only judge the correctness of received information but also its usefulness (Requirement 2). And similar to the use cases with defects, the behaviour of the cooperation partner is time-varying (Requirement 7). There are disruptions in the partner's behavioural pattern.

9. A group of ten cars reports bad recommendations regarding another car, which are wrong.

   This is a lying witnesses use case (Ramchurn et al., 2004; Sen and Dutta, 2002). Such an attack is possibly easy to perform, but can have a high impact on the reputation system. This use case is often mentioned in the context of trust and reputation. For this reason, I mention it here, although it addresses the reputation mechanism, not the individual-level trust mechanism. So it is out of the scope of this dissertation.

   The trust mechanism can help the reputation system by evaluating received recommendations. A recommendation is also a kind of report (Requirement 2).

   Altogether, if a car can well judge about acts of cooperation, own experiences – if there are enough – should dominate the trust in another car over the other's reputation. This stabilises the reputation system, too. In the course of time, the trusting system can learn the balance between own experiences and various kinds of recommendations.

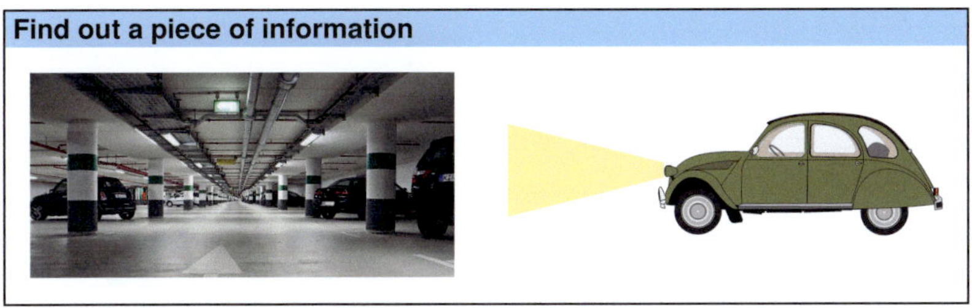

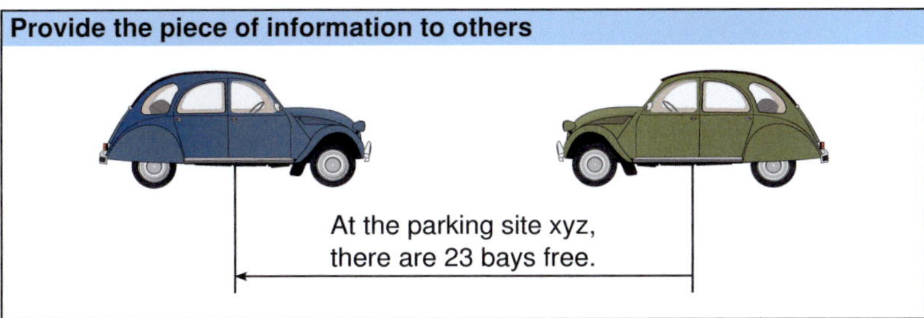

Figure 8.1: The trust-related situation as it occurs in a typical vehicular network scenario: the information retrieval happened some time in the past before the information exchange.

8.3 The Trust Problem

The use cases describe the car in a situation where it must decide about its future behaviour. This decision may be influenced by several considerations: Is the sender trustworthy? Is the received information plausible with regard to the own knowledge? What is the benefit and cost if the information is accepted correctly and what if it is accepted wrongly? The decision may be even more complex, if one vehicle reports congestion ahead and another vehicle mentions congestion on the alternative route. This thesis focuses on the correctness of the received message only. This is the trust-related aspect of the decision problem. All other aspects are out of scope.

To analyse the trust problem, two related scenarios are compared: the typical scenario of information exchange in a vehicular network as shown in Figure 8.1 and a fictitious scenario with the same aim but different steps as illustrated in Figure 8.2.

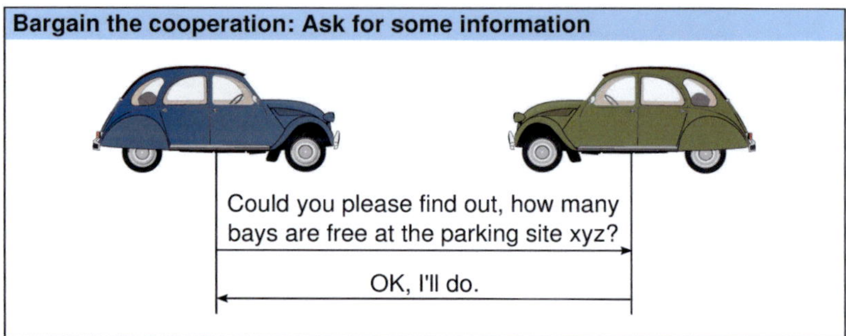

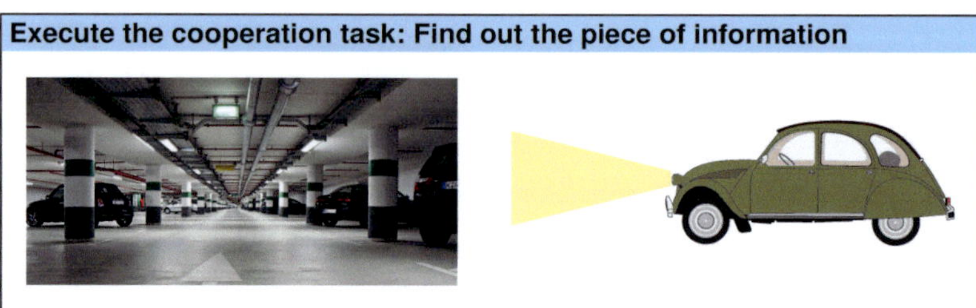

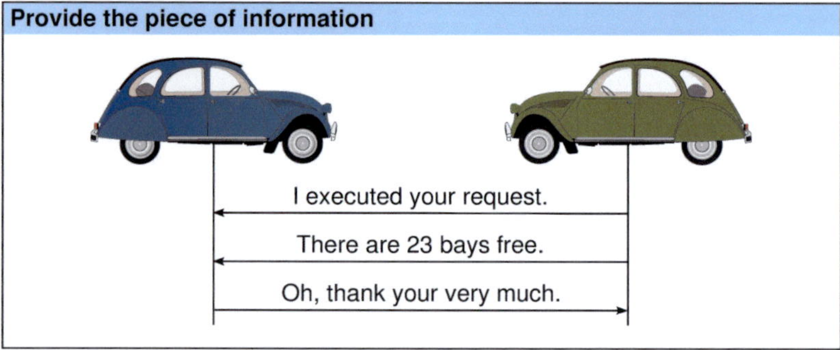

Figure 8.2: The trust-related situation as it occurs in a fictitious scenario: first the bargaining of the cooperation, then the information retrieval and finally the information exchange.

In the considered trust situation, the vehicle wants to know something, so it needs cooperation. In the fictitious scenario, it requests cooperation explicitly by sending a request. Then the other vehicle cooperates, if it answers with a correct value. This procedure corresponds to what people typically connect with trust and delegation: request and bargain the delegation, execute it and hand out the result. It is illustrated in Figure 8.2. In a vehicular network, the other vehicle usually sends some values without a previous request though. It just broadcasts information to all passing vehicles (Figure 8.1). There is an implicit agreement that the spread values should be correct and relevant. Thus an implicit request can be imagined before the automatic spreading of information. When taking this into account, receiving a message means that cooperation has implicitly been requested and the other vehicle has an answer.

But the trusting vehicle may still not know, whether the other vehicle cooperated. This is so, because it may not know, whether the received message is correct. In some cases, it can verify the value only at a later point in time. Until then, the vehicle can believe the message or not. If it does, it accepts the costs of a possible betrayal. This is a key feature of a trust situation. Only with this choice, the trustor really opts for a trustful act of cooperation with the other vehicle.

If the trusting vehicle accepts the message, it cooperates. The subject of cooperation is information about a property of the world. As mentioned above, the bargaining is often implicit, describing a request for correct and useful information. Consequently the bargained cooperation outcome is the correct value of this property; and the real cooperation outcome is the received message. Thus in this trust setting, the real cooperation outcome is known, while the implicitly bargained outcome is wanted. This is different to other cooperation scenarios and can be seen best by comparing Figures 8.1 and 8.2.

In summary, the trust computation can be formalised as an inference problem. The trusting vehicle wants to infer the unknown bargained outcome $\mathcal{B}_O$ (the correct piece of information) from the real cooperation outcome $\mathcal{T}_O$ (the received message) and all other evidence e mentioned in the Enfident Model. This means the probability distribution $P(\mathcal{B}_O \mid \mathcal{T}_O, e)$ should be computed. Section 8.6 shows how to do this based on the Chapters 6 and 7.

Note that a vehicle cannot verify every value it receives with own observations. Either the regarded location is not on the route of the vehicle or the vehicle takes a new route because of the received information (e.g. about a congestion). In some cases though, the vehicle can build its opinion from the values of many other vehicles. Thus it does not need to observe everything on its own. In some other cases, like the answer "I don't know", it may be that the vehicle can never know whether the other vehicle really did not know the answer or whether it refused to cooperate.

This leads to another problem of trust computation here: The cooperation outcome is verified with what the trusting vehicle assumes to be right. And this assumed truth depends in turn on information from other acts of cooperation. In the consequence, trusting only works if there are either many own observations or many good reports.

The sections of this chapter up to here introduced the vehicular network scenario in general. The following sections show how it is used to evaluate the Enfident Model through simulation.

8.4 Simulation Environment

This section gives an overview of the evaluation software. It highlights especially the parts I developed. Figure 8.3 illustrates the interaction of the tools.

To virtually rebuild the scenario of cooperating cognitive vehicles, traffic is necessary. The vehicle movements of the traffic determine when a vehicle can talk with which other vehicles. Thus they constitute something like the social structure of the vehicles. The movements are generated with *SUMO* (Behrisch et al., 2011), a microscopic traffic simulator of the German aerospace centre (Deutsches Zentrum für Luft- und Raumfahrt, DLR), and with own extensions. Section 8.4.1 introduces the concept of the social structure and the tool chain.

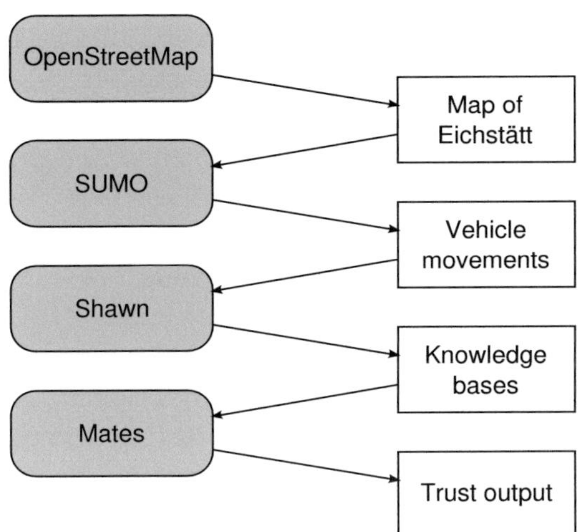

Figure 8.3: The tool chain used for the simulation. The boxes on the left represent the tools, those on the right the output/input files. SUMO is a traffic simulator, Shawn a communication network simulator, and Mates an own application for post-processing, which can apply various trust models.

The movements of the vehicles are then passed to the network simulator *Shawn* (Kröller et al., 2005). It provides the infrastructure to let the vehicles communicate. I extended Shawn by a new plug-in (named *vanet*) for the whole information processing in the vehicles. Sections 8.4.2 and 8.4.3 introduce the processing in this new Shawn plug-in.

Finally I developed a new application called *Mates* for some post-processing. This application can take knowledge bases from a Shawn simulation and inject synthetic message patterns (see Section 8.4.4). It can then replay these modified knowledge bases with various trust algorithms.

8.4.1 The Social Structure

The development of trust depends on interaction characteristics like when cooperation happens, how often it happens, or for what purpose it happens. For this evaluation, I summarise these characteristics under the term *social structure.* As the social structure determines the trust development, it must be included in an evaluation method for trust-enabled systems. This is the reason, why I decided to evaluate the trust algorithms with the scenario of cognitive vehicles that cooperate in a vehicular network. This scenario contains a social structure that can be reproduced well.

In the scenario, the traffic forms the social structure. It describes the movements of the vehicles and, thus, determines their communication partners. Because the drivers take some routes regularly (like to work or to a specific shopping mall), some vehicles meet each other again and again. The resulting meeting pattern constitutes the base for the trust development.

But how can such traffic be created? It is insufficient to generate the traffic completely randomly. The resulting vehicle movements would not reflect the described meeting pattern. Instead I developed a simplified activity-based traffic demand model. Such a demand model creates the traffic based on statistical characteristics of the population, like work place densities in every street, number of children per family or the unemployment rate. Activity-based traffic demand modelling is well described in Hertkorn, 2004. Traffic demand created in this way is random but includes the required inner dependencies. Vehicles can meet again and again like work colleagues or friends.

The traffic simulation has been performed with the microscopic traffic simulator SUMO, a comprehensive open source traffic simulator. The activity-based demand model has been implemented as an additional module of SUMO. The map for the traffic simulation is based on a cut of the German town Eichstätt from OpenStreetMap (OpenStreetMap Project, 2013). The simulation replays the trips of 6554 vehicles over 18 weeks and results in a detailed description of their movements.

These movements are then put in the sensor network simulator Shawn to simulate the communication between the vehicles. This simulator can process a sensor network with a large number of nodes quickly (Kröller et al., 2005). Shawn implements various models of wireless networks. The evaluation in this thesis uses a simple disk model with a range of 100 m; the transmission is reliable. This simplification is necessary, because the main purpose of the simulation here is the evaluation of a trust algorithm. For this,

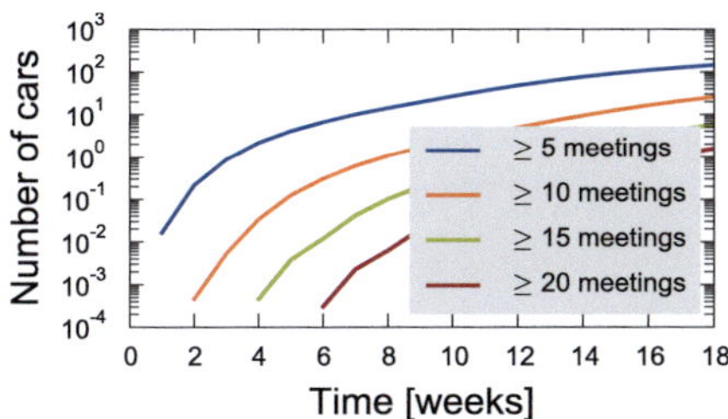

Figure 8.4: This graph shows how often a prototypical vehicle meets other vehicles. For example, on average a vehicle meets about three other vehicles at least once a week. Each line filters the set of considered senders based on the number of messages received from that sender. The x-axis shows the simulated time in weeks, the y-axis the number of senders, from which so many messages have been received.

it is important that an error in a received message can be attributed solely to the sender's behaviour.

Figure 8.4 visualises the resulting social structure. It shows, for example, that a prototypical vehicle can exchange information with about three other cars at least once a week on average. Thus a vehicle has something like well-known friends. To generate the figure, every vehicle in the simulation counts the number of received messages per sender, cumulated over the whole simulation. After every week, the sender-receivers pairs are assigned to bins according to their count. A sender from which twelve messages were received so far is assigned to the bins "$\geq$ 5 meetings" and "$\geq$ 10 meetings". Then the number of sender-receiver pairs per bin is counted and divided by the number of receivers. The resulting values are drawn in the figure, where every kind of line represents a bin.

8.4.2 Events and the Information Model

So far, the vehicles just exchange simple messages without content. To simulate the information processing in the vehicles, I developed the new plug-in *vanet* for the network simulator Shawn. It is part of the Shawn project and thus open source.

Type of the event	Number of events	Mean of the duration	Standard deviation of the duration
Short-term	400	1.5 hours	0.75 hours
Medium-term	80	4 days	2 days
Long-term	8	2 years	1 year

Table 8.1: The parameters of the event generator.

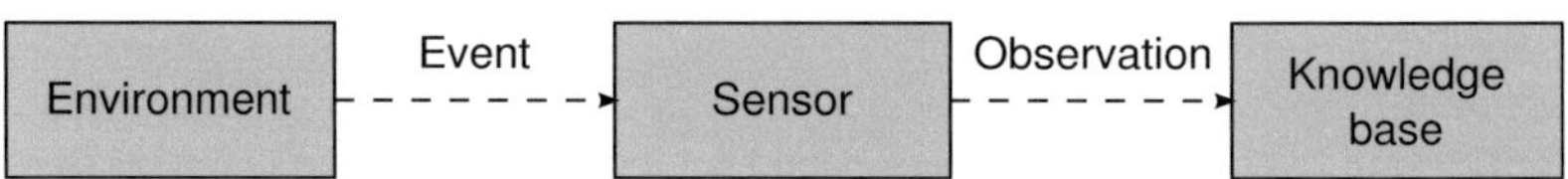

Figure 8.5: The flow of information from the environment into a vehicle's knowledge base. The sensors represent the gate, through which information gets into the vehicular network.

To induce information the vehicles could talk about, environmental events are placed in the map. They can be perceived by a vehicle's virtual sensor. The events are of three different types, for long-, medium- and short-term information. These types correspond to the classes in Chapter 8.1. The value of an event is discrete in the range 0–4. The events are randomly generated on the map. Their values are distributed equally over the value range. Events exist only for a limited time. Their beginning of life is distributed equally throughout the simulation; their duration of life is drawn from a Gaussian distribution with a mean and variance depending on the information type. Table 8.1 lists the generator's parameters. All events are stored in a configuration file and read by the vanet plug-in at simulation start-up.

8.4.3 The Processing in the Vehicles

The trust model should help the vehicle to obtain correct pieces of knowledge. In the simulation, other vehicles may send wrong values either because of sensor errors or because of a manipulated transmission module.

New information gets into the vehicular network through the sensors of the vehicles (see Figure 8.5). Every vehicle has one sensor. If a sensor finds an event in its environment, it adds a normally distributed error to the

Sensor type	Standard deviation for medium-term events	Standard deviation for all other events
S0	0.00	0.00
S1	0.67	0.33
S2	0.33	0.67
S3	1.00	1.00

Table 8.2: Sensor types and the standard deviations of the additive Gaussian noise that is applied to the sensor values. Note that the type S0 has a deterministic error of 0, which is indicated in the table by $\sigma = 0$.

perceived value and reports the modified value to the vehicle. There are four sensor types. The sensor type defines the parameters of the Gaussian noise. It is always mean-free. Its standard deviation depends on the type of the perceived event and is listed in Table 8.2. The noisy values are rounded to integers and limited to the range 0 to 4. Figure 8.6 on the next page shows an approximate distribution for the resulting error. It applies the right rounding. Just the range limiting cannot be reproduced correctly, because it depends on the actual sensor value the noise is added to. A perceived event is called an *observation*. It is saved in the vehicle's knowledge base for later transmission in the network.

In the vanet plug-in, a vehicle simply spreads out the latest observations (see Figure 8.7 on the following page). Some vehicles, however, are configured to disseminate wrong information. They take the latest observations from the knowledge base, change their values to zero and send the manipulated values. This manipulating disseminator should represent defects or malicious software in the vehicles.

A piece of information received from another vehicle is called a *report* in this document. It is also saved in the knowledge base as Figure 8.8 on the next page illustrates.

A vehicle can derive its own opinion about an event from the corresponding observations and reports. This process is usually called information integration or data fusion (see e.g. Nakamura et al., 2007). It is the place where trust comes in. Reports from a highly trusted source should have a higher

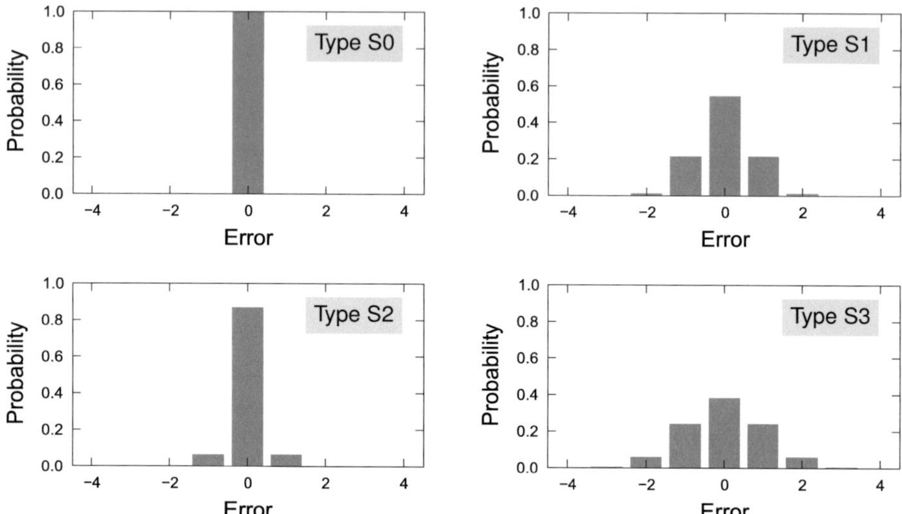

Figure 8.6: Approximate distributions of the sensor error for medium-term events. These figures show the error distributions, when samples of the normal distributions in Table 8.2 are rounded to integers and when those samples outside the range {-4, ..., 4} are mapped to their nearest range member.

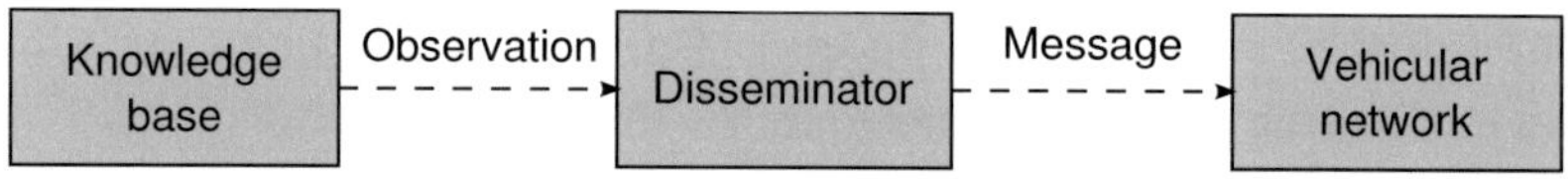

Figure 8.7: The flow of information in the vehicular network. This is the way how vehicles spread out what they have perceived from the environment.

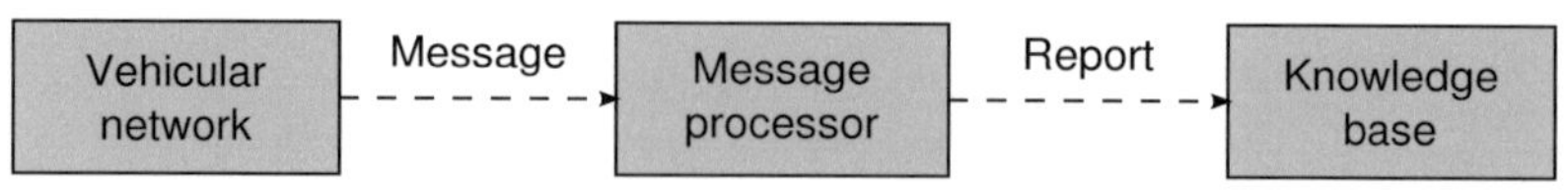

Figure 8.8: The flow of information that comes from the vehicular network. This is the way, how a vehicle stores a cooperation history in the knowledge base.

influence on the resulting piece of knowledge than reports from a weakly trusted source. Thus a trust-supported integration algorithm is necessary. The algorithm used for the evaluation is very simple. It has been introduced first in Bamberger et al., 2012.

The vehicle then uses its own opinion to judge about the reliability of received reports. The absolute value of the difference between the own opinion and the report's value quantifies the cooperation outcome in the trust model.

Finally a vehicle drops redundant reports from the knowledge base. If it finds more than five reports from the same sender and about the same event in its knowledge base, it removes some of them in a way that the time of observation is evenly distributed between the reports with the lowest and the highest time stamp. This process happens regularly. It ensures that the size of the knowledge base keeps reasonably. Otherwise the simulation would get very slow and finally run out of memory. Tests showed that this forgetting algorithm has little impact on the simulation result.

Altogether the knowledge base is the hub for the information processing in the vehicle. It is connected to other vehicles and environmental events through task-specific internal components. In addition, it saves the history of cooperation with other vehicles. From this history, the vehicle derives its assumed true value of an event and the error of a cooperation outcome.

8.4.4 Post-processing with Mates

The combination of SUMO and Shawn reproduces the real world in a form suitable for this thesis. To investigate a trust algorithm, additional synthetic patterns of received messages are needed though. For example, to analyse the generalisation and specialisation capabilities of an algorithm, it would be helpful that a vehicle sees the exact same message pattern several times throughout the simulation (see Section 8.5.2). The program *Mates* can modify a knowledge base in the needed ways. Mates is a tool, which I wrote to support various actions around the simulation. It is especially important for the following two actions.

Mates can inject reports in knowledge bases that exhibit a certain pattern. This functionality is used in the sensor quality scenario to investigate generalisation and specialisation, and in the defect scenario to analyse the response of a trust algorithm to time-varying behaviour.

Moreover Mates can replay knowledge bases to evaluate a trust algorithm. For the algorithm, this looks like a real run in a vehicle. During the replaying, Mates records the development of trust in selected senders. At the end, it assesses the final trust for all sender-received pairs. The resulting trust values and distributions are the basis for the diagrams in the next chapter.

Replaying a knowledge base with Mates is different from a simulation with Shawn though. Shawn only saves the 25 largest knowledge bases, not all. This is necessary to speed up the replaying. Even for only those 25 knowledge bases, replaying takes two to six weeks.

8.4.5 Further Limitations of the Simulation Environment

A simulation only considers a limited image of the reality. This image must be designed to be sufficient for the evaluation purpose. Most of the simplifications of the presented evaluation method were already described above. This section points to some other limitations that should be kept in mind, but may be obscure so far.

The activity-based traffic demand includes the work place activity only. The vehicles drive from home to work and back. This is a very simplified demand model. In reality, vehicles would drive many more trips. This would lead to a denser traffic. Consequently a vehicle would meet more other vehicles and could cooperate more often. This simplification has the advantage that it results in a better manageable simulation complexity, while it is sufficient to highlight the main properties of the Enfident Model.

Chapter 1.1 already points out that a trusting system must be able to assess all relevant facets of a cooperation outcome. For this reason, all information in the network comes from a vehicle's sensor; and the sensor of all vehicles can perceive the same kind of information. Roadside units, radio broadcasting and other information sources are omitted. The simulation considers car-to-car communication only, with cars that are autonomous regarding the information flow (not necessarily regarding the driving).

Furthermore the simulation tools run independently. As a result, the driver who receives a message about a traffic jam ahead cannot change the route. This example is a use case, in which a vehicle would not be able to verify the correctness of the received message by an own observation, because it would avoid the traffic jam. Verification could be supported by other means like central traffic reports though. I do not consider such scenarios here, mainly because the coupling between SUMO and Shawn (which already exists) slows down both programs heavily. The simulation of this thesis only realises information flows that can be verified autonomously and that do not affect the driving route.

8.5 Simulation Scenarios

The simulation environment described so far is used in different configurations: the simulation scenarios. This section introduces them and shows what evaluation purpose they suit for. They are all based on a run of the traffic simulator SUMO and the network simulator Shawn. The resulting knowledge bases are then modified with Mates to contain some additional message sequences. Finally these knowledge bases are replayed with the trust algorithms mentioned in Section 8.6.

8.5.1 Verifying the Fulfilment of the Requirements

The evaluation aims to show that the proposed implementation of the Enfident Model meets the requirements for a trust model as postulated in Chapter 4. This way of verification closes the loop that started with the use cases and the requirements and, up to here, ended with the implementation. The simulation scenarios are like test cases for this verification. The following list shows, how every requirement is verified.

Requirement 1. The sensor quality scenario (Section 8.5.2) is suitable to show whether the computed expectation indeed depends on the selected influences. It features the ability-related aspect of trust.

Requirement 2. The Enfident Model represents trust in the form described in this requirement. Trust values and trust distributions appear throughout the evaluation results. This shows their usefulness.

Requirement 3. The Enfident Model separates trust and decision making as required. No additional verification is necessary.

Requirement 4. The Enfident Model learns from experiences as demanded in this requirement. All evaluation scenarios are based on experience-based learning. The effect of present constraints is not evaluated though.

Requirement 5. The proposed implementation of the Enfident Model works on the complete knowledge base and reconsiders a previous judgement with each received message. However outside the trust algorithm, the knowledge management realises forgetting to handle the high amount of data (see Section 8.4.3). As a result, the vehicle can only revise previous judgements based on the reduced data. Because the requirement is partially violated outside the trust model, it is not further evaluated here. The theory already shows that the Enfident Model basically fulfils the requirement.

Requirement 6. The sensor quality scenario (Section 8.5.2) can show how a trust algorithm learns the behaviour of a certain partner step-by-step with each act of cooperation (specialisation). And it can also demonstrate how an algorithm learns for all partners at once (generalisation).

Requirement 7. The defect scenario (Section 8.5.3) is suitable to show how a trust algorithm handles time-varying behaviour of the cooperation partner.

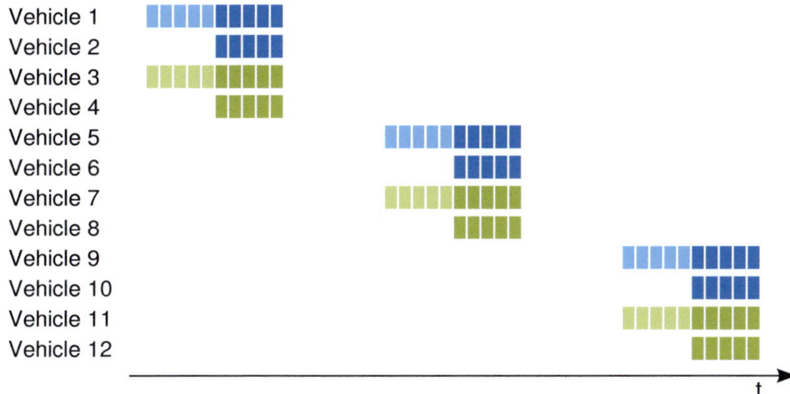

Figure 8.9: Schema of the reports inserted in the knowledge bases after the Shawn simulation. The horizontal axis indicates the progressing simulation time; the vertical axis gives the name of the sending vehicle. The coloured blocks represent sequences of reports with the following content:

Information type: long-term, values: 1, 1, 1, 0, 1 (sensor type 1)
Information type: medium-term, values: 1, 0, 0, 2, 1 (sensor type 1)
Information type: long-term, values: 1, 1, 1, 1, 1 (sensor type 0)
Information type: medium-term, values: 1, 1, 1, 1, 1 (sensor type 0)

8.5.2 Sensor Quality Scenario

The sensor quality scenario aims to show, whether a trust algorithm can learn a model of a partner's ability-related behaviour. In the simulation environment, a vehicle's ability-related behaviour corresponds to its sensor type as given in Table 8.2 on page 177. The quality of the sensor should be the only reason, why there is wrong information in the network. Only own observations are spread out; and no manipulating disseminator is used. Under these conditions, the behavioural model that is estimated by the trust algorithm matches with the sender's sensor model and can thus be verified. This scenario is similar to Use Case 1 and is suitable to verify Requirement 1.

Because a vehicle continuously meets new other vehicles, this scenario can also show how a trust algorithm handles new cooperation partners. So it reproduces the Use Case 2 as well. This aspect can be better evaluated, if the new cooperation partners are comparable. For this reason, two times three new vehicles are inserted in this scenario with Mates. They send comparable message sequences at three points of the simulation. Figure 8.9 on the previous page illustrates schematically what reports are added. At the beginning of the simulation, vehicle 2, which has a sensor of type S1, sends a certain pattern of messages with medium-term information. The vehicles 6 and 10 repeat the same message sequence in the middle and at the end of the simulation. The vehicles 4, 8 and 12 do the same, but have a sensor of type S0. So the vehicles 2, 6, and 10 and the vehicles 4, 8 and 12 are directly comparable. They make it possible to see, how a trust algorithm changes its treatment of new cooperation partners over time.

With the insertions, the scenario is also suitable to investigate the generalisation and specialisation capabilities of a trust algorithm (Requirement 6). The judgement about a new partner might correspond to a generalisation over all known partners. And with every other message received from a new partner, the algorithm should quickly be able to judge the partner's individual trustworthiness.

Generalising over cooperation partners does not only mean to learn appropriate initial trust. Rather a trust algorithm should also find inner dependencies between the various abilities. In the simulation environment, it is possible to infer from the standard deviation for the long-term information type to that for the medium-term information type. The reader can see that easily from Table 8.2. To show this form of generalisation, another six vehicles are inserted: the vehicles 1, 5 and 9 send the same message sequence as the vehicles 2, 6, and 10, but with five messages with long-term information in advance. So they perform a different task prior to the task that the vehicles 2, 6, and 10 also do. The trust algorithm should be able to exploit this additional information to better judge about the messages with medium-term information. In the same way, the vehicles 3, 7 and 11 replicate the vehicles 4, 8 and 12 just with some preceding messages of the long-term type.

All in all, the inserted reports follow the competence schema of the preceding Shawn simulation. They do not violate the idea of this scenario as described in the beginning of this section. They just help to get meaningful data out of the simulation regarding the Requirement 6.

To complete the description of the scenario and to make it reproducible, I list the configuration of Shawn here.

```
# Make the randomness of the simulation reproducible
random_seed action=set seed=1

# Set up the world
prepare_world edge_model=simple comm_model=disk_graph range=100

# Configure logging (logging needs a world)
logging_load_cfg log_cfg_file=logger.cfg

# Move nodes by a SUMO dump file.
node_movement mode=sumo processors=vehicle \
              net_file=eichstaett.net.xml \
              dump_file=eichstaett.dmp.xml

# Set up the reading for the vehicle sensor
create_environment readings_file=eichstaett.readings.xml

# Set up the tasks for the statistics output
output_interval=604800

# Configure the information processing
sensor_error=medium_mixed_errors
trust_computer=bamberger2010
prune=overfull

init_vanet

# Run the simulation for 18 weeks (time in seconds)
simulation max_iterations=10886400 run_without_nodes=true
```

8.5.3 Defect Scenario

This scenario should show, how a trust algorithm can react to changes in the partner's behaviour. It addresses Requirement 7. Its idea is adapted from Use Cases 5 and 8. The simulations with SUMO and Shawn are the same

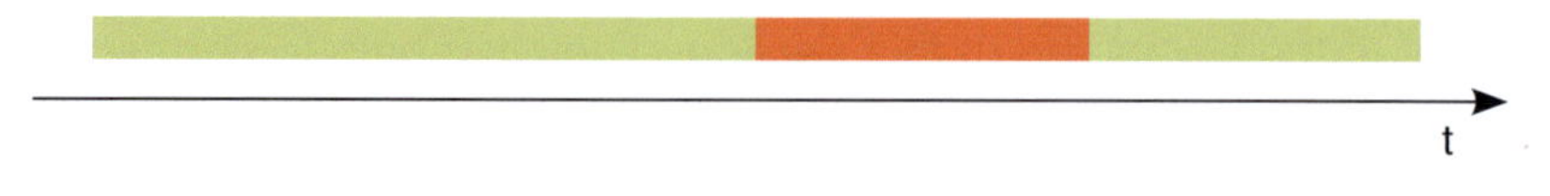

Figure 8.10: Schema of the reports inserted after the Shawn simulation for the defect scenario. The horizontal axis indicates the progressing simulation time. The green areas represent correct messages; the sending vehicle is working well. The red area in the middle marks a period, in which the vehicle is defect and, thus, disseminates wrong messages.

as in the sensor quality scenario. Then the knowledge bases are modified with Mates in the following way.

Four vehicles with sensors of type S0 are inserted. They send many good messages during about the first half of the simulation time. Then they become defect and, thus, send wrong information. The value in the message is always zero. After about three quarters of the simulation time, they get repaired and send correct values again. Figure 8.10 illustrates this schema for one of the inserted vehicles.

In summary, a vehicle's reliability depends on the quality of its sensor and its defect state in this scenario. This means for a trust algorithm that it must handle two alternations in the other's behaviour for the inserted vehicles.

8.6 The Algorithms Used for the Evaluation

The evaluation uses the Credit algorithm and variations of the Enfident Model. The Credit algorithm has been selected, because it is a typical representative of today's trust algorithms and is based on some elaborate work in Ramchurn, 2004. It is defined there completely. This section introduces the evaluated realisations of the Enfident Model.

To realise the Enfident Model in the presented scenarios, various attributes could be identified and assigned to the three entity classes. This would result in a trust algorithm, which is too complex for the scope of this evaluation. Instead I designed a model, which is complex enough to capture all features to be evaluated, and simple enough too lead to interpretable results. Figure 8.11 depicts the basic relational setting. In the simulation, a receiver can identify the sender by the sender id in the message. Further the

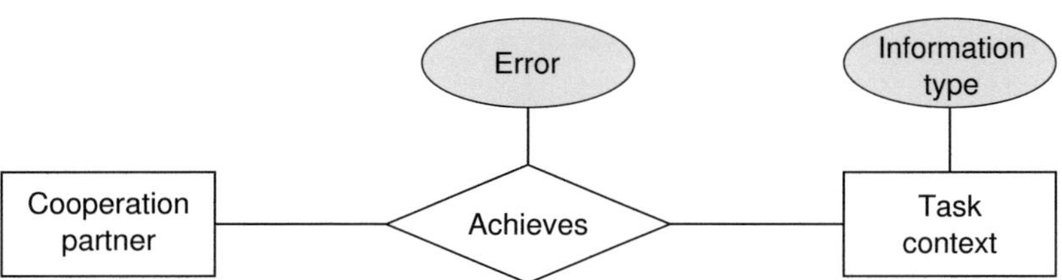

Figure 8.11: Basic relational model for the evaluation setting. It is widely reduced compared to Figure 6.3 to match the purpose of the evaluation.

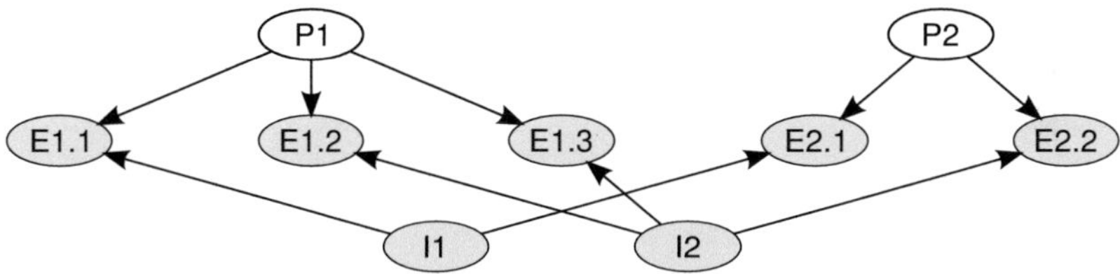

Figure 8.12: Exemplary Bayesian network for the Enfident-sp-ot algorithm. It contains a hidden variable (P1, P2) for each cooperation partner as well as the observed message errors (E1.1, . . . , E2.2) and the observed information type (I1 = L, I2 = M).

message contains the information type, which describes the task to perform. And the value in the message is transformed in an absolute error.

This basic data model is converted in four trust algorithms. They are described in the following list. The corresponding figures illustrate all the same example. It consists of two senders, P1 and P2. From P1, one message was received with long-term information (L) and two messages with medium-term information (M). From P2, one message is available for each of both information types. The somewhat technical names are just to distinguish them in later references. sp stands for static partner, dp for dynamic partner, ot for observable task and ht for hidden task.

Enfident-sp-ot. This is the simplest realisation. Figure 8.12 illustrates it as a Bayesian network for the above example. One hidden variable represents each sender. It is sampled as described in Section 7.2.2. The message error is the only child attribute. The information type is

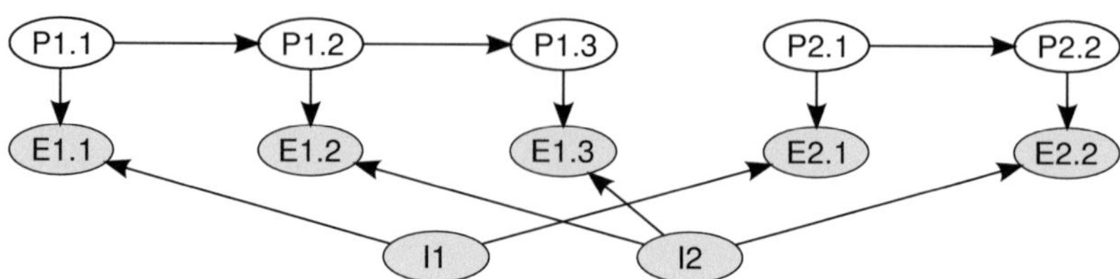

Figure 8.13: Exemplary Bayesian network for the Enfident-dp-ot algorithm. It contains a hidden variable (P1.1, ..., P2.1) for each interaction with the cooperation partners. The observed message errors (E1.1, ..., E2.2) depend on the corresponding partner variables and the observed information type (I1 = L, I2 = M).

taken as the constant value of the task context. Thus the task context is an observable variable in this algorithm. The message error also depends on this variable.

Enfident-dp-ot. For this algorithm, the partner is modelled as a time-varying process following Section 7.2.3. The task context is still realised as an observable constant: the information type (see Figure 8.13).

Enfident-sp-ht. This algorithm equals with the Enfident-sp-ot algorithm, but the task context is represented as a hidden variable as described in Section 7.2.2. Two messages that regard the same event (same information type and same location) share the same hidden variable for the task context. The task context T3 in Figure 8.14 shows these connections. The information type is an observable child attribute of the task context.

Enfident-dp-ht. This variant combines the time-varying model for the cooperation partner (Section 7.2.3) with the time-invariant hidden variable model for the task context (Section 7.2.2). Thus it combines the modifications of the Enfident-dp-ot and the Enfident-sp-ht algorithm as depicted in Figure 8.15.

To keep the figures clean, the distribution parameters are omitted. As a result, the task contexts in Figures 8.14 and 8.15 seem independent from

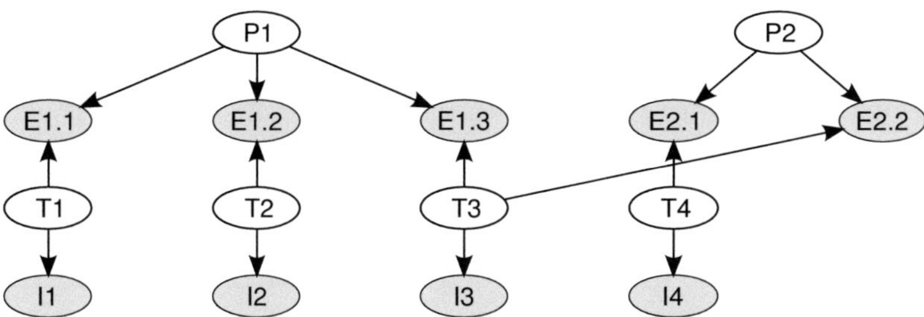

Figure 8.14: Exemplary Bayesian network for the Enfident-sp-ht algorithm. It contains a hidden variable (P1, P2) for each cooperation partner as well as the observed message errors (E1.1, ..., E2.2). In addition, hidden variables (T1, ..., T4) represent the different task contexts. The information type variables (I1 = I4 = L, I2 = I3 = M) describe the task contexts.

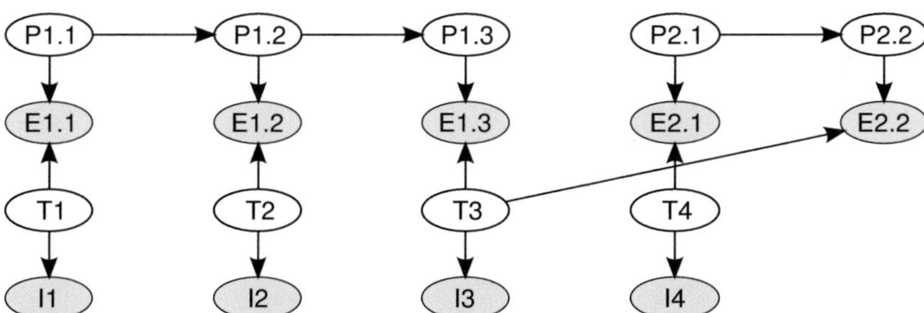

Figure 8.15: Exemplary Bayesian network for the Enfident-dp-ht algorithm. It contains a hidden variable (P1.1, ..., P2.1) for each interaction with the cooperation partners. Additional hidden variables (T1, ..., T4) represent the different task contexts. The observed message errors (E1.1, ..., E2.2) and the observed information type variables (I1 = I4 = L, I2 = I3 = M) are associated with the corresponding hidden variables.

each other. This is not the case though. They are connected through the parameters and the values of the information type variable.

The evaluation results in the next chapter have mostly been produced with the Enfident-dp-ot algorithm. It realises all necessary features and is still simple enough to have well interpretable results. The other variants help to contrast the effects of different implementation techniques. Note that I finally removed the algorithms with infinite-dimensional mixtures from the evaluation results. Their additional flexibility seems relevant for real, more complex scenarios. In the presented, simple evaluation scenarios, they did not add something interesting as tests showed. But they need much more computation time and memory.

9 Evaluation Results and Their Discussion

To show that the Enfident Model meets the requirements of Chapter 4, simulations have been performed as described in the previous chapter. The current chapter presents and discusses the simulation results. It tests whether the requirements and the Enfident Model lead to sensible and intuitive results in some well defined scenarios. And it verifies the Enfident Model with regard to the requirements (see Section 8.5.1).

Section 9.1 shows, how the evaluated algorithms learn competence-related trust from past interactions (Requirements 1 and 4). In addition, it presents trust values and trust distributions side by side to show their respective usefulness (Requirement 2). Section 9.2 looks closer at the trust development to visualise specialisation and generalisation (Requirements 4 and 6). The response to time-varying behaviour is shown in Section 9.3 (Requirement 7). Finally Section 9.4 points out that the purpose of trust directly corresponds to a common problem of learning.

9.1 Learning the Competence-Related Influences

The results in this section are based on a simulation of the sensor quality scenario as introduced in Section 8.5.2. At the end of the simulation, the 20 largest knowledge bases are taken. For all senders in these knowledge bases, trust is assessed in the following way: For each sender, a fictitious new report of that sender is passed to the trust algorithm. This is done for both, medium-term and long-term information types and provides task-specific trust distributions and trust values. These results suit to see whether

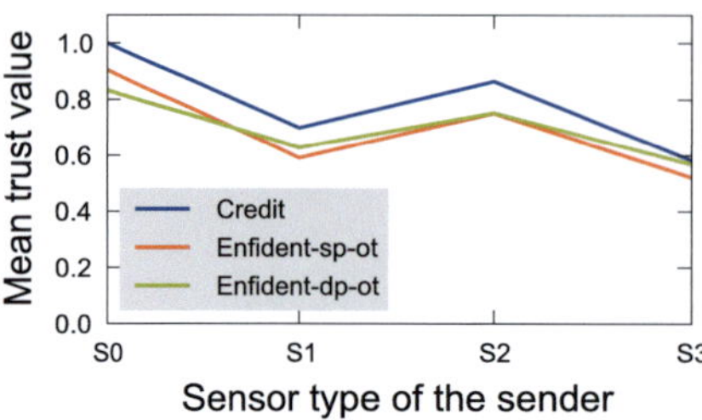
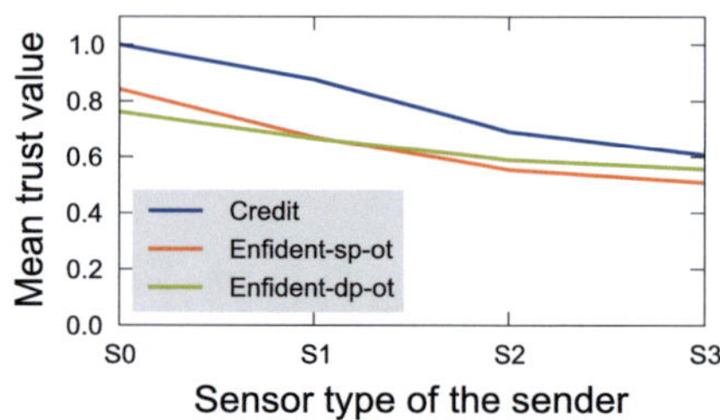

Figure 9.1: The evaluated trust algorithms can well learn the competence of a cooperation partner. The figure shows the mean trust values depending on the type of the sender's sensor and the trust algorithm. Only senders, from whom at least six message have been received, are included. The diagram on the left hand side is based on fictitious reports about medium-term information, that on the right hand side about long-term information.

a trust algorithm can learn the competence of the cooperation partner. Note that the resulting trust is associated with a new fictitious report not with the message that was actually received from the sender. This makes the results interpretable because they are independent from other messages about the same subject.

Only senders, from which at least six messages have been received for a given information type, are considered in the following. This is done because few single reports of a sender cannot fully specify its behaviour – they are just random. The resulting trust distributions and trust values are visualised in the following two ways.

The trust value as a reliability indicator. The trust values are grouped according to the type of the sender's sensor. For every group, a mean value is computed. Figure 9.1 shows the means. The sensor type of the sender is indicated on the x-axis, the mean of the trust values on the y-axis and the trust algorithm by the line style. The trust values of the diagram on the left hand side are based on the fictitious reports with medium-term information; those on the right hand side come from the fictitious reports with long-term information.

In the sensor quality scenario, a vehicle has sensors with an error variance depending on the information type. Table 8.2 on page 177 lists them. If trust values are considered a reliability measure, the table suggests that higher variances should lead to lower trust values. And since the variances depend on the information type and the sensor type, the trust values should do so as well. This is what Requirement 1 says. Figure 9.1 shows that all evaluated trust algorithms indeed compute trust values related to the error variance. Thus a receiving vehicle can estimate the reliability of the sender well after a couple of messages. For the Credit algorithm, the designer of the implementation must define the grouping according to the information type explicitly though. In contrast, the algorithms based on the Enfident Model find the groups on their own by clustering as long as sufficient attributes are provided. Note that only the relative scaling of the trust values is important. The absolute scaling can easily be adjusted with a non-linear transformation as, for example, shown in Bamberger et al., 2012.

Trust as a mechanism to model the other's behaviour. For Figure 9.2 on the following page, the trust distributions of the Enfident-dp-ot algorithm in response to a fictitious medium-term report are taken and grouped according to the sender's sensor type. Then a box plot is drawn for every group. The x-axis shows the absolute value of the error, the y-axis indicates the estimated probability of the error. The grey boxes indicate the range between the lower and upper quartile, the orange line the median. The grey whiskers illustrate the values in a range that is extended at the lower and upper end of the boxes by one and a half of the interquartile range. All outliers outside this extended range are drawn as single crosses. The additional green circles mark the actual error distribution of the data in all twenty knowledge bases. This is the reference distribution. Since the Credit algorithm cannot provide a trust distribution, the figure reproduces the trust distributions only for the Enfident-dp-ot algorithm.

Note that the green reference distribution, which is based on the actually transferred messages, differs from the Gaussian noise distribution of the sensor error model. The uncorrected standard deviations of the refer-

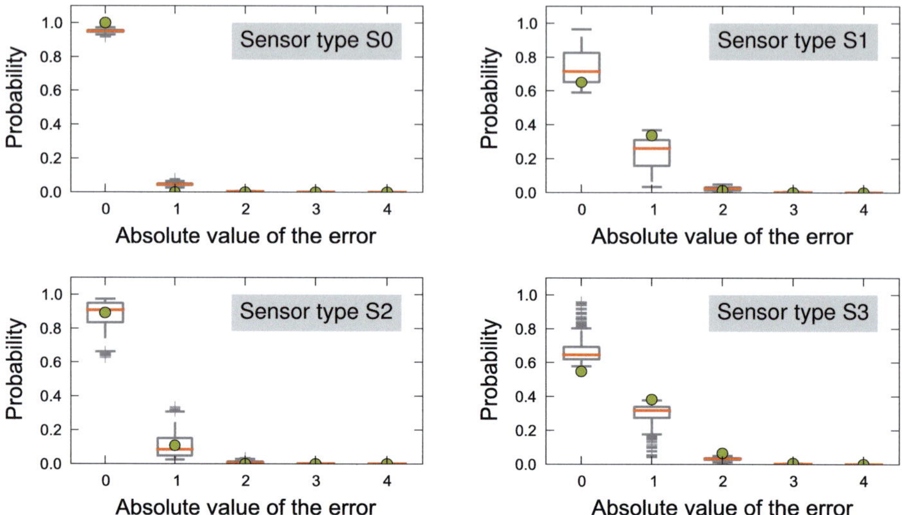

Figure 9.2: The Enfident-dp-ot algorithm successfully learns Gaussian distributions as behavioural models. The figures illustrate the outcome distributions in response to a medium-term message depending on the sender's sensor type. They are box plots with orange medians as well as grey whiskers and outliers. The reference distribution is marked with green circles (compare it with Figure 8.6 on page 178).

ence distributions in Figure 9.2 – given a mean error of 0 – are 0.00, 0.62, 0.33 and 0.83 for the sensor types S0, S1, S2 and S3 respectively. They correspond to the standard deviations of the sensor error model in the middle column of Table 8.2, but are smaller than those. The deviation from the sensor error model is higher, if the value of the standard deviation is higher. This comes from the post-processing of the sensor values in the simulation. Particularly the error can be ±4 at most. Higher errors, which would lead to higher standard deviations, cannot occur. Thus the error model is indeed not completely Gaussian, but quantised and range-restricted.

If a trust algorithm should learn a model of the other's behaviour, the trust distribution, which is computed for a report in the given simulation setting, should reflect the error distribution of the originating sensor. The reason for this is that a vehicle only sends own observations in the simulation. The figure illustrates the trust distributions. The used implementation of the Enfident Model does not assume a Gaussian model, but uses a categorical distribution, which induces few assumptions. It just learns the sender's reliability and finds distributions with a Gaussian-like shape similar to the error distribution of the originating sensor. Thus it meets its purpose fulfilling Requirements 1 and 4.

Note that the time-varying Enfident-dp-ot algorithm judges good sensors a bit too bad and bad sensors too good. This is not the case for the time-invariant Enfident-sp-ot algorithm. Section 9.4 discusses this phenomenon further.

Representation of trust. Figures 9.1 and 9.2 are based on the same data. They show the same information, one in form of trust distributions, the other one in form of trust values, which are derived from the trust distributions. This justifies parts of Requirement 2; and this illustrates that both representations of trust can go hand in hand with the Enfident Model as recommend in Section 4.3.1.

Discussion of the results. The simulation results show that the presented trust algorithms can learn the reliability of the sender's sensors well. But isn't the setting too simple with a behaviour, which depends just on the information type and the sender, and without a complex information integration algorithm in the sender? Yes, the simulation scenario is simplifying. And some tests showed that the amount of erroneous messages significantly decreases, if the vehicles send values that come from the integration of several information sources. The reason for letting vehicles send just their own observations is that a sender's behaviour should have a probabilistic model, which is known in advance to make the simulation results interpretable. In contrast, the proposed trust algorithms do not assume anything about what sender's behaviour constitutes. The behaviour may depend on the sensor, the information integration algorithms, manipulation software or anything else. As a consequence, the simulation results suggest that the Enfident-dp-ot algorithm can evaluate behaviour that depends on various factors within the sender (Requirement 1).

Moreover, the clustering of the Enfident-dp-ot algorithm finds the dependency on the sender and the information type automatically. It is just me, the investigator, who has the pre-knowledge about it to interpret the simulation results correctly. Thus the trust algorithm might also handle various other influencing attributes for clustering. This is in contrast to many other trust algorithms in the literature like the Credit algorithm. For them, the algorithm designer has to define categories for trust in advance (see Chapter 4).

All in all, this section features a competence-dependent and time-invariant behaviour of the cooperation partner. Especially Figure 9.2 shows that the algorithms that are based on the Enfident Model can learn the opponent's behaviour without many prior assumptions. This is what a trust mechanism should do. The resulting trust distributions can then support the data fusion algorithm to prioritise good information sources (see Section 9.5).

9.2 Specialisation and Generalisation

This section looks at the vehicles that are added to the sensor quality scenario through post-processing with Mates. They are described in Section 8.5.2 and Figure 8.9 on page 183. These vehicles are suitable to show, how the trust algorithms develop specific trust over several interactions and how they learn to judge new situations.

A trust algorithm can specialise for a certain situation with more and more experiences for that situation. For example, a vehicle could receive several messages from the same sender and about the same kind of subject in similar contexts over time. Regarding the Enfident Model, this means to learn the joint outcome distribution for fixed entities. And this is what most experience-based trust algorithms do (see Section 4.4). It is shown on an example below.

A trust algorithm can also *specialise* on some entities only, while the others vary. For example, a vehicle can interact with a certain cooperation partner in several different tasks. This way, it gets to know more and more facets of the other's overall behaviour: It learns the other's competence and willingness. The same can be expressed the other way around: A trust algorithm can *generalise* over some entities only, while others remain fixed. In the previous example, the trusting vehicle can exploit similarities between the different tasks to estimate the cooperation outcome of situations, for which no or only few experiences with the given partner are available. This partial specialisation or generalisation is shown in two forms below.

The data set used in this section is extensive. Therefore I first introduce it, before I derive the key results. Finally I discuss specialisation and generalisation at the end of this section.

Description of the data. The common frame of the results in this section is the following: After every message of a vehicle that has been added during post-processing, a dummy report is presented to the trust algorithm. The trust value for this report is taken as the current trust value. The dummy report comes from the same sender as the preceding message, but is not associated with a real event of the simulation – no other vehicle can have sent a message about the same subject. This ensures that the dummy report is always treated in the same way. So the trust values are associated with this dummy report, not with the messages that are illustrated in Figure 8.9. The diagrams in this section are all from the perspective of the vehicle 3038; it is the receiver and the trustor. To describe the results compactly, only trust values are shown; the trust distributions are omitted.

Figure 9.3 on the following page shows the resulting trust values from the Credit algorithm for vehicles 9–12. The reader can imagine the time running from left to right. All messages that are vertically in line arrived at the same simulation step; the order of processing within a simulation step is unknown. The diagrams on the left are associated with the long-term messages, those on the right with the medium-term messages of Figure 8.9. The same evaluation has been performed with the Enfident-dp-ot algorithm. Figure 9.4 on page 199 shows it.

With vehicles 9 to 12, the previous paragraph considered the response of the trust algorithm at the end of the simulation; the algorithm can access

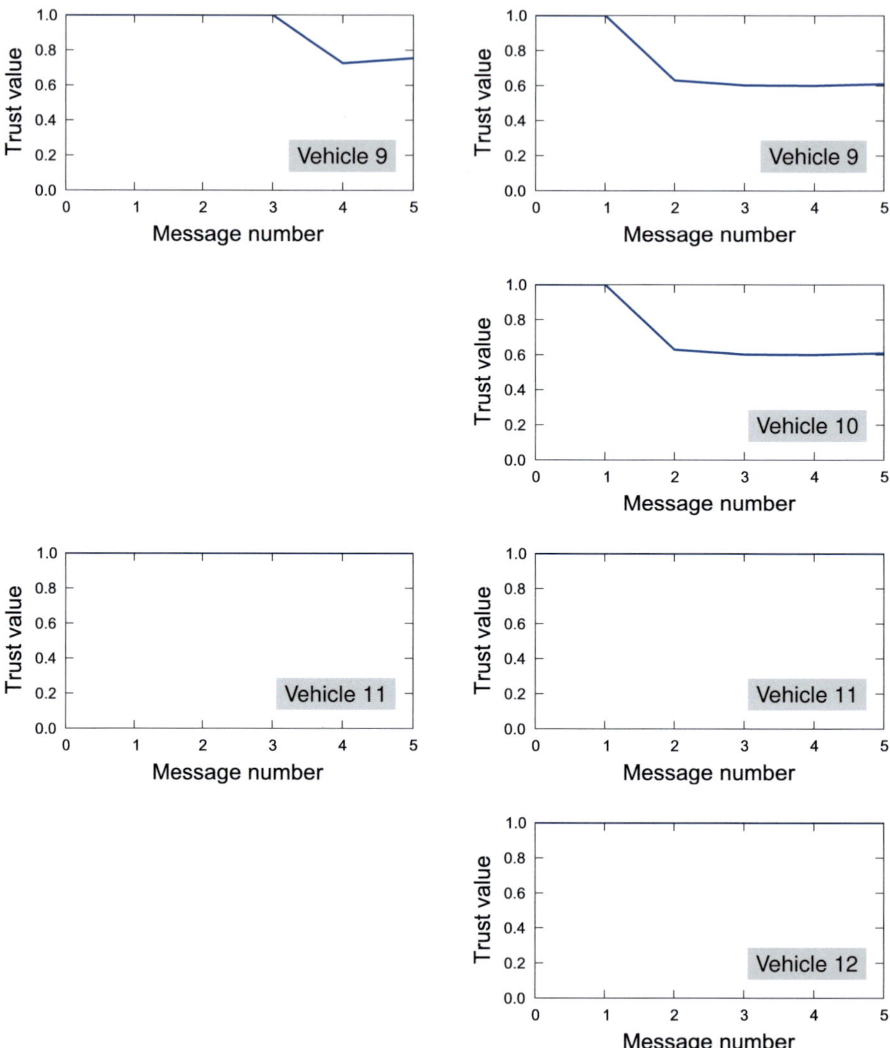

Figure 9.3: Specialisation with an increasing amount of experiences and transfer learning from one task to another with the Credit algorithm. A single vehicle (3038) receives all messages of vehicles 9 to 12. The temporal order runs from left to right. Vehicles 9 and 11 start with five messages containing long-term information. The right column regards messages with medium-term information. The Credit algorithm specialises on the sender with an increasing amount of experiences, but it cannot generalise from one type of message to another one.

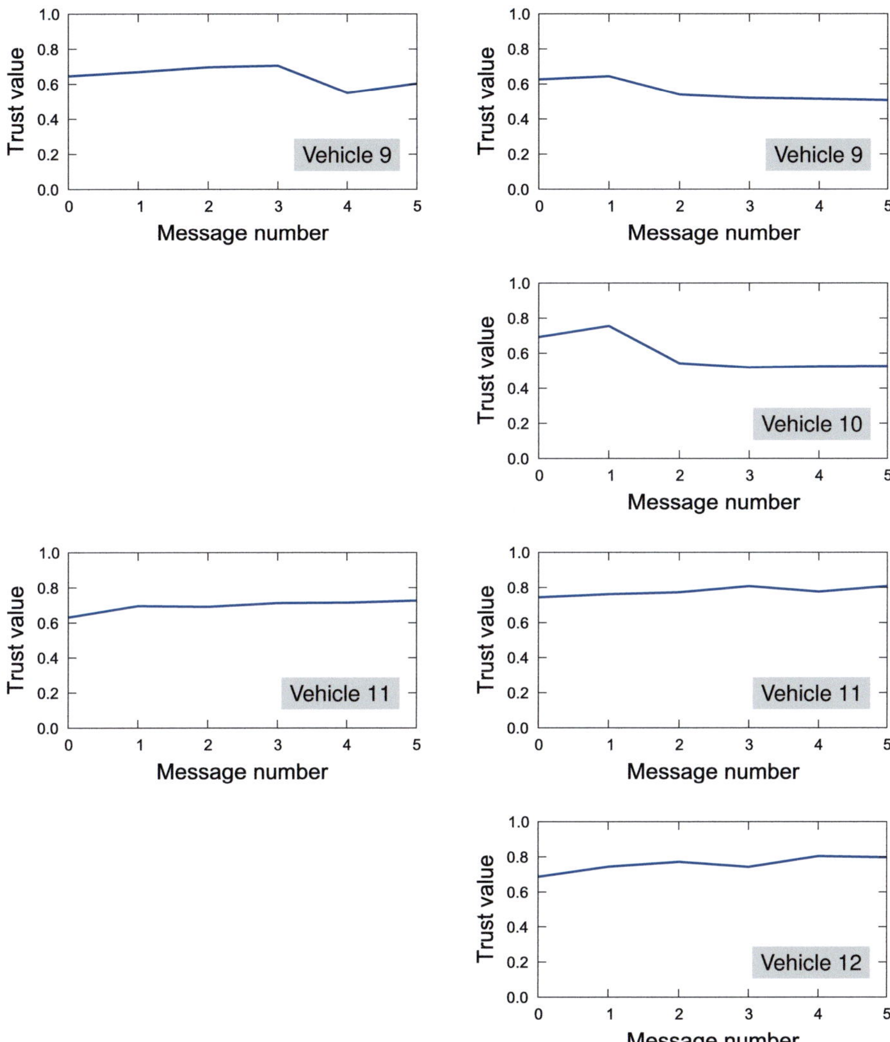

Figure 9.4: Specialisation with an increasing amount of experiences and transfer learning from one task to another with the Enfident-dp-ot algorithm. A single vehicle (3038) receives all messages of vehicles 9 to 12. The temporal order runs from left to right. Vehicles 9 and 11 start with five messages containing long-term information. The right column regards messages with medium-term information. The Enfident-dp-ot algorithm specialises on the sender with an increasing amount of experiences; and it generalises from one type of message to another one, as the initial trust value for messages with medium-term information depends on the preceding messages with long-term information

much evidence at this time. In contrast, Figure 9.5 shows how the Enfident-dp-ot algorithm evolves with more and more overall experiences, from the beginning and the middle of the simulation until its end. Vehicles 2, 6, and 10 send the same message sequence, just at different simulation times. The same applies to vehicles 4, 8, and 12. In Figure 9.5, these different points in time are ordered from top to bottom. The left diagrams are associated with vehicles that have a sensor of type S0, the right diagrams with those that have a sensor of type S1. Note that the diagrams for vehicles 10 and 12 are the same as in Figure 9.4. Also note that this evaluation has been omitted for the Credit algorithm, because its behaviour regarding new vehicles does not change over time. The diagrams are always the same as those of vehicles 10 and 12 in Figure 9.3.

What do all these results say about generalisation and specialisation? I consider two cases in the following.

Full specialisation.   If the same act of cooperation is performed again and again (in the same contexts), a trust algorithm can learn the underlying statistical process – the joint outcome distributions for the participating entities. Both, the Credit algorithm and the Enfident-dp-ot algorithm, can do that. For example, the diagram of vehicle 12 in Figure 9.4 shows that the algorithm learns the high reliability (sensor type S0) of that vehicle message by message: The line is increasing. In contrast for vehicle 10 with sensor type S1, the diagram shows a decreasing trust value. The algorithm learns that the sender is not as good as supposed in the beginning. Figure 9.3 shows similar results for the Credit algorithm. Thus both trust algorithms learn stepwise to judge the reliability that comes along with a certain trust situation. They exploit the available experiences (Requirement 4) to learn the influence of certain entities (Requirement 6).

Transferring knowledge between task contexts.   Vehicles 9 and 10 send the exact same sequence of medium-term information as Figure 8.9 visualises. For this reason, the Credit algorithm assigns the same trust values to those vehicles (Figure 9.3). However vehicle 9 sends some long-term

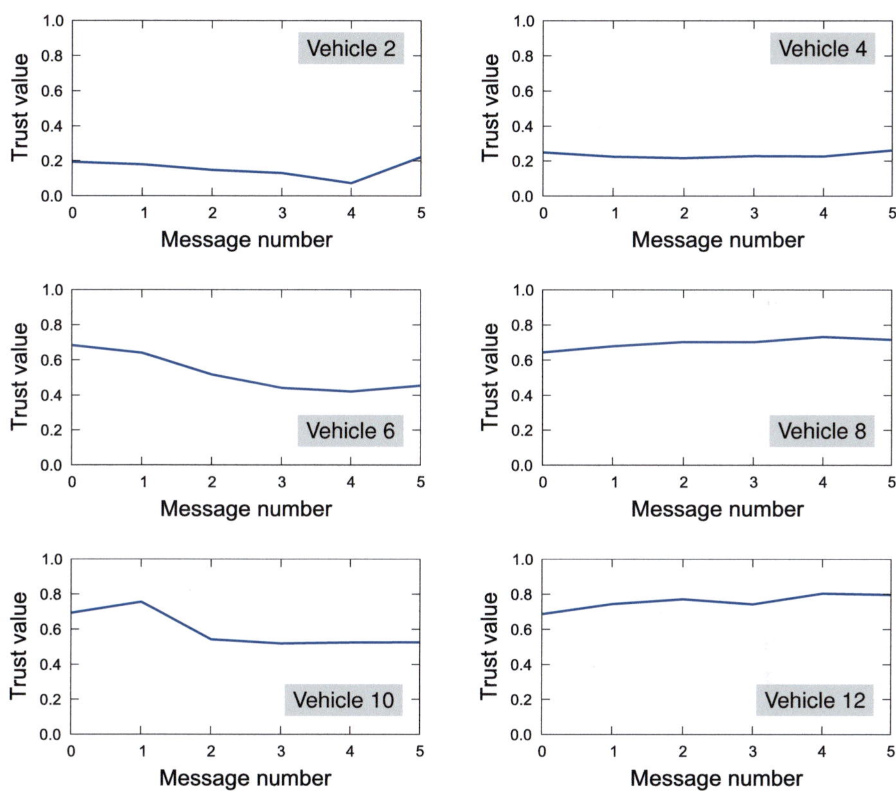

Figure 9.5: The Enfident-dp-ot algorithm can learn initial trust by generalising over all entities. The figures show how vehicle 3038 rates the five medium-term messages received from vehicles 2, 4, 6, 8, 10 and 12. The two figures on the top come from messages at the beginning of the simulation, those below from the middle of the simulation and the two at the bottom from the end of the simulation. The senders of the diagrams on the left have bad sensors for medium-term information (sensor type S1); those on the right have good sensors (sensor type S0). The initial trust value is learned with an increasing amount of experience in the receiver (from top to bottom).

messages in advance. Figure 9.4 shows that the Enfident-dp-ot algorithm can take advantage of them. The Enfident-dp-ot algorithm learns from the preceding long-term messages that vehicle 9 is less reliable than it would initially assume without the long-term messages. Consequently it starts with lower trust values for the medium-term messages of that sender compared to the trust values for vehicle 10. This means that the Enfident-dp-ot algorithm can learn the kind of partner from experiences in one task context and transfer this knowledge to another context with the same partner. The same effect can be observed with vehicles 11 and 12. This time, the Enfident-dp-ot algorithm learns from the long-term messages that vehicle 11 is more reliable than the average unknown vehicle. So the graph for medium-term information of vehicle 11 begins higher than that of vehicle 12. In contrast, the Credit algorithm cannot generalise. The graphs for the medium-term messages of vehicles 9 and 10 as well as that of vehicles 11 and 12 are exactly the same.

Transferring knowledge between cooperation partners. The Enfident-dp-ot algorithm can learn the initial trust in a new cooperation partner. It can develop different initial trust depending on the task context. This means the algorithm generalises over the partner entities, while it specialises on the task context entity. This can be seen in Figure 9.5. It shows only those vehicles that only send the five messages with medium-term information. The initial trust value increases on both sides with every row, that is, with an increasing amount of experiences. The Enfident-dp-ot algorithm learns the performance of a typical unknown cooperation partner. In the beginning, it knows nothing about the error statistics. Then it learns that high errors are less likely than low errors. Consequently the trust value increases. The initial trust value also includes that half of the vehicles in the simulation have a sensor of type S0 and, thus, are very reliable. Starting from this initial trust value, the algorithm then specialises on the specific partner: On the left, the trust values increase, while they decrease on the right. Just at the beginning of the simulation (the first row), the Enfident-dp-ot algorithm cannot yet judge the other's trust reliably. The curves remain vague. In contrast, the Credit algorithm always starts with a trust value of one before any message has been evaluated. It cannot adapt its initial trust to the trustworthiness of all the cooperation partners.

Discussion of the results. Specialising takes evidence from the entities, which are participating in the trust situation of interest, in order to learn the joint probability distribution of those entities. In contrast, generalising takes the evidence from other entities to learn that joint probability distribution. It regards transfer learning, which is important for the trust mechanism to handle new situations and, thus, to act under uncertainty. The evaluation results illustrate this. They justify Requirement 6 and make clear, why this requirement is inherent to trust. Rettinger et al. (2008) emphasise the importance of transfer learning as well. Only few trust algorithms address this ability at present though (see Chapter 4.4.2).

Specialisation and generalisation are antagonists, which complement each other in two ways as the evaluation results show: First, if a vehicle knows one of the trust situation's entities, it can specialise on that entity and generalise over the other entities using the experiences, which have been made with that entity. Second, a vehicle can use specialisation and generalisation for one single entity at the same time. It can take all experiences, which have been made with that entity, and complement them with the experiences, which have been made with all the other entities of the same root type exploiting similarities. While the Enfident Model connects both forms, other trust algorithms may support only the one or the other. For example, the trust algorithm in Bamberger et al., 2012 can only generalise over different tasks. The experiences that are used to compute trust are restricted to those with the same cooperation partner.

In some scenarios, the cooperation partner cannot be identified reliably. Then the algorithm cannot specialise on the partner entity. Vehicular networks can be one example of such a scenario, as pseudonyms will possibly be used there to protect privacy. For this and other reason, Raya (2009) considered traditional trust models inappropriate for vehicular networks and proposed to focus on the plausibility of the data with a fixed logic. She calls this form of trust data-centric trust. The evaluation results show though that the Enfident Model can still specialise on all other entities, even if the cooperation partners do not interact again and again or do not recognise each other. This means that the Enfident Model implicitly includes data-centric trust. It automatically moves its attention on the cooperation partner or the data, just as needed. Thus the mixed specialisation and generalisation is important to keep in mind, when analysing various trust applications.

9.3 Time-varying Behaviour of the Trustee

The vehicles that have been considered so far always follow the same statistical process throughout the simulation. This section investigates how three trust algorithms handle changing statistical processes. It is based on the defect scenario as described in Chapter 8.5.3 and looks at the vehicles that were added through post-processing with Mates. These vehicles change their underlying processes from a reliable partner to a defective partner and back to a reliable partner as illustrated in Figure 8.10 on page 186.

Description of the data.   The way, how the trust value develops with every received message, can well describe how an algorithm adapts to the statistical processes of the considered senders. Figure 9.6 shows this development for two senders: vehicle 1 on the left and vehicle 2 on the right. In the first phase, all messages contain correct values. In the second phase, both vehicles are defective. They always send 0. The first wrong message of vehicle 1 is the 33rd, that of vehicle 2 is the 30th. In the third phase, the defect is repaired. The vehicles send correct messages again. The first correct messages of vehicles 1 and 2 are the 44th messages. The top row of the figure shows the error in the received message. This error is the input of the trust algorithms. Note that the vehicles do not necessarily send wrong values. Rather they always send 0, which may accidentally be true or false. The receiver is always the same vehicle (3038). Three different trust algorithms have been simulated: the Credit algorithm, the time-invariant Enfident-spot algorithm and the time-varying Enfident-dp-ot algorithm. The results are given row by row in the figure as indicated by the labels in the diagrams.

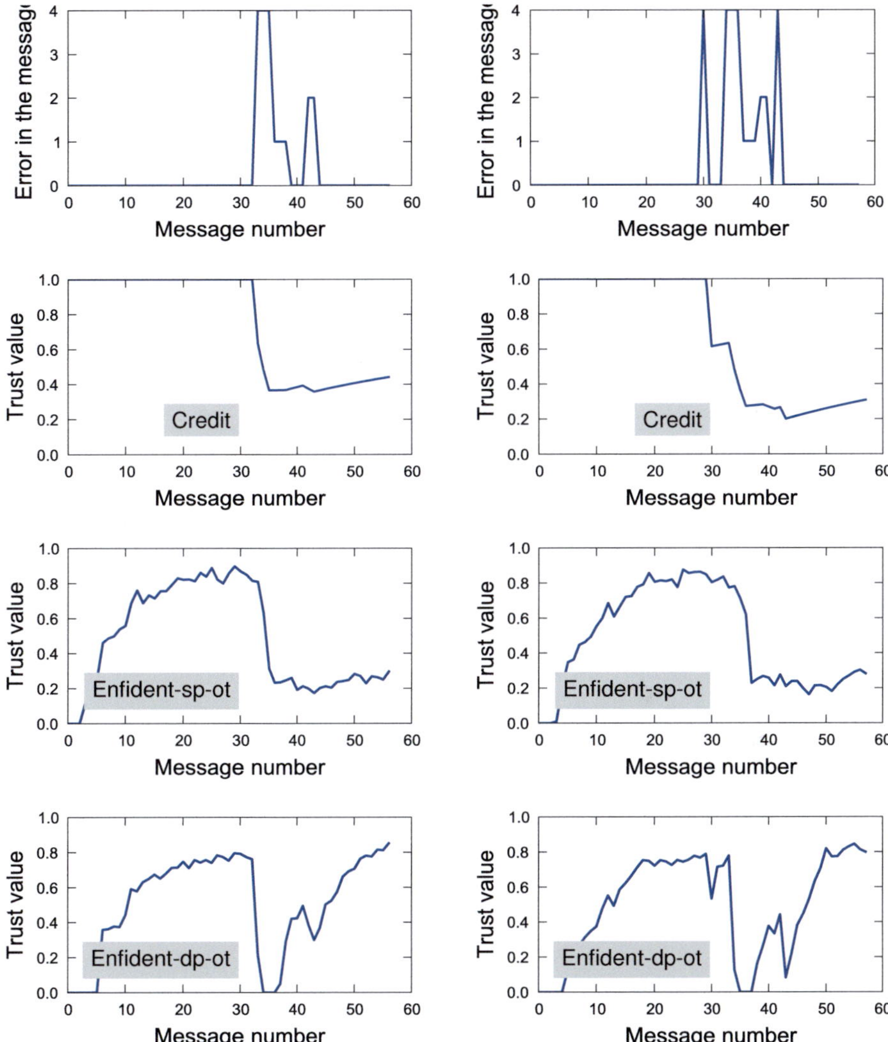

Figure 9.6: Response of the trust model on a time-varying behaviour of the cooperation partner. On the left, the sender is vehicle 1; on the right, it is vehicle 2. The first row shows the error in the received message. It describes the behaviour of the sender. This is what the trust algorithms see. The following three lines illustrate the responses of three different trust algorithms in the receiver. The name of the algorithm is indicated in the diagram.

Adaptation to a time-varying process.   When looking at the curves before the 30th message, all trust algorithms judge both senders as very reliable. With the sequence of wrong messages, the trust values are going down quickly. They stay around 0.2 to 0.4 during the defect phase. This is common to all three algorithms. But then a major difference becomes visible. The Credit algorithm and the Enfident-sp-ot algorithm can react to the next sequence of correct messages only slowly. The trust value increases only by around 0.1. In contrast, the Enfident-dp-ot algorithm ends with a trust value of about 0.80 to 0.85. This trust level is comparable to that before the defect phase. So the Enfident-dp-ot algorithm can follow a time-varying process. In contrast, both other algorithms treat the sender as a single process. The algorithms take all previous outcomes disregarding their temporal order. As a consequence, they weight the bad message in the third phase, as if they occurred directly before the current message.

The different approaches of the algorithms can also be seen at the beginning of the second phase. There the effect is not as clear-cut as in the third phase though. The Credit and the Enfident-sp-ot algorithm take the high errors in the beginning of the second phase (value four on the y-axis) together with the low errors in the first phase as part of a Gaussian-like distribution. The lower trust values reflect just the overall performance of the sender, not a switch in the estimated statistical process. Consequently both algorithms tend to a medium to low trust value. In contrast, at the beginning of the second phase, the Enfident-dp-ot algorithm estimates a trust distribution that has high probabilities especially at errors of zero and four. The algorithm goes up to a medium trust value only with the following medium errors in the messages. Then it accepts a Gaussian-like distribution. In the consequence, the later errors of four lead to a smaller peak down, as they fit better in the newly estimated process. This can be seen in the right diagram of the Enfident-dp-ot algorithm.

Discussion of the results.   The plots for the Credit and the Enfident-sp-ot algorithm may seem right at first glance. A cooperation partner who betrayed once looses his trustworthiness. Trust shall recover slowly. However all three algorithms do not distinguish between increasing and decreasing trust. The longer the first phase with the correct messages takes, the slower trust will decrease with the time-invariant trust algorithms. This is in contrast to an algorithm that considers the order of messages like the Enfident-dp-ot algorithm does.

Bad experiences can be emphasised with other tools like utility functions and the mapping from a trust distribution to a trust value. Indeed the latter mapping of all three algorithms is quite pessimistic. For this reason, the time-invariant algorithms decrease their trust values rather steep, although the trust distribution does not change much. The increase is the slower though.

In the literature, only few trust algorithms address abrupt changes in the partner's behaviour as Section 4.4.3 points out. Most algorithms focus on time-invariant or slowly changing behaviour. The proposed model for time-varying entities leads to good results for time-invariant and time-varying behaviour of cooperation partners, as the various results in this chapter. The actual performance depends on the concrete implementation algorithm though.

9.4 Generality of a Trust Algorithm and Its Convergence

A trust algorithm should learn the behavioural model of a cooperation partner so that it can predict cooperation outcomes. The possible behavioural models of any partner may not be known in advance though. This is in contrast to a sensor model. The sensor is known and can be tested. As a consequence, a Gaussian model, for example, can be valid for a sensor, but could be too restricted for the behaviour of a cooperation partner.

This shows that a trust algorithm should be able to capture a variety of different behavioural models at the same time. It should be generic. To emphasise this, I selected categorical distributions for the error model in this evaluation. They can take on any form for a discrete magnitude. Moreover

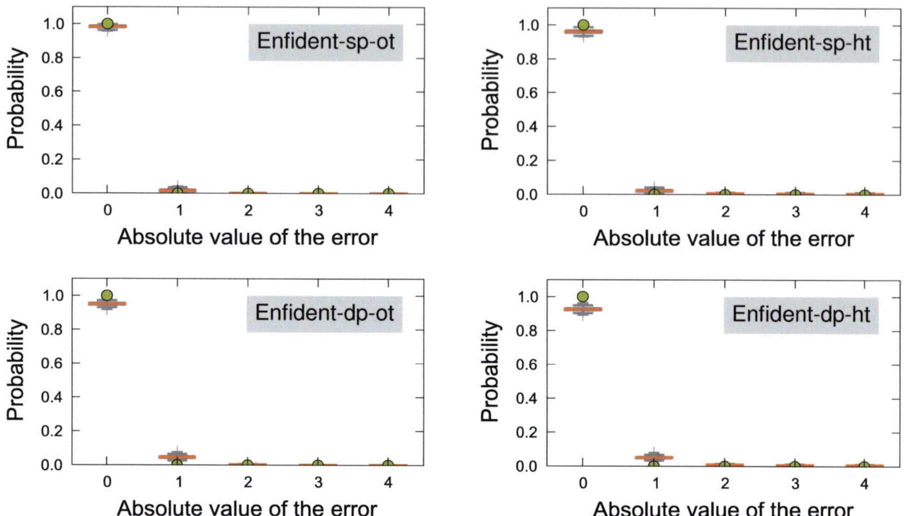

Figure 9.7: More specific algorithms find the right explanation of the data better than more generic algorithms. This figure shows box plots of the found behavioural model for senders with a sensor of type S0. These plots correspond to the top left diagram in Figure 9.2. The plots only differ in the trust algorithms as indicated in each diagram.

the time-varying algorithms are more generic than the time-invariant algorithms. They can learn even more forms of behaviour.

On the downside, a more generic algorithm has more ways to explain the experiences. In the consequence, it may need more data to converge to the real explanation. Figures 9.7 and 9.8 illustrate this. The diagrams are box plots like in Figure 9.2. The bottom left diagram of Figure 9.7 is the same as the top left diagram of Figure 9.2, and the bottom left diagram of Figure 9.8 is the same as the bottom right diagram of Figure 9.2.

Figure 9.7 depicts the outcome distributions, which the different algorithmic variants from Section 8.6 learn for senders with a sensor of type S0. This sensor type is an extreme case, as it makes no error. The most restricted algorithm, the Enfident-sp-ot algorithm, converges best to the real distribution (the green circles). In contrast, the Enfident-dp-ht algorithm has more degrees of freedom. Consequently it has more ways to explain the

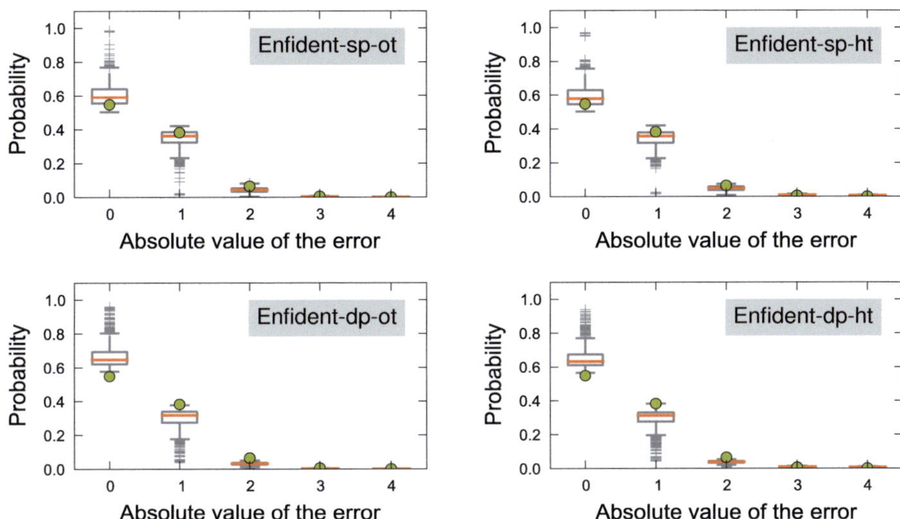

Figure 9.8: More specific algorithms find the right explanation of the data better than more generic algorithms. This figure shows box plots of the found behavioural model for senders with a sensor of type S3. These plots correspond to the bottom right diagram in Figure 9.2. The plots only differ in the trust algorithms as indicated in each diagram.

same data and these ways are more pessimistic. It would need much more data to rule out the pessimistic explanations. The other two algorithms are somewhere in between.

The other extreme case is the sensor type S3. The outcome distributions for senders with such a sensor are shown in Figure 9.8. Again the Enfident-sp-ot algorithm gets close to the real distribution of the data, while the Enfident-dp-ht algorithm finds other more optimistic explanations of the same data.

In summary, trust development requires a generic underlying model, which can capture various behavioural patterns of the cooperation partners. Such generic algorithms may need much data to converge to the real behavioural pattern. At a first glance, this need contradicts the purpose of the trust mechanism to support acting under uncertainty, which means acting in case of little data. There seem to be contrary needs in Requirement 6.

But in fact, an algorithm can offer both: It can give reasonable advice based on none or little data for a single cooperation partner; and it can specialise on various different behavioural patterns. The challenge is to generalise with few overall data, which comes from other partners than the cooperation partner of interest, and to specialise to the behavioural pattern of that specific partner with little additional data. Consequently the algorithm designer must typically balance the need for data and the generality of the underlying model.

For completeness, I must note here that the proposed algorithms fail to balance both needs, because the scenario is quite simple. The statistical model with categorical distributions suits well to emphasise that the trust algorithm can find Gaussian shaped distributions without prior knowledge about the Gaussian sensor model. But the statistical model uses indeed mixture models, each of which is a superposition of categorical distributions (see Equation 7.1 on page 120). If the mixture model of the error variable, for example, contains K mixture components and every component is a categorical distribution with five probabilities, four of which are independent, then the resulting mixture distribution has $4K$ independent parameters. But its shape only has four independent values. This consideration changes though, if more than one variable depends on the mixture component. Other distributions like discretised normal distributions are useful for the mixture components as well.

9.5 Summary

The evaluation results visualise in various forms, what a statistical model of the cooperation partner's behaviour means here. Section 9.1 considers a single trust situation at a given point in time. The shown trust distributions predict the other's behaviour for this situational context. The evaluation demonstrates that with task-dependent trust distributions. A trust value sums up a trust distribution on a one-dimensional scale. As a consequence, it misses some information of the trust distribution. The evaluation did not consider attributes of the bargaining context though. Because the Enfident

Model is symmetric regarding its entity classes, the development of trust depending on such attributes would work in the same way as with the task attributes. The additional situational constraints of Requirement 4 were also omitted to keep the evaluation focused. They could easily be integrated in the sampling, as well, but add only little to the understanding of trust.

The evaluation especially emphasises the temporal view on trust development. Sections 9.2 and 9.3 show, how the behavioural model (the trust distribution) is learned for a specific combination of attributes; how the initial trust in a new cooperation partner evolves with an increasing amount of overall experiences; how the experience with a certain partner for one task can be exploited to estimate the cooperation outcome with the same partner but for another task; and why a trust algorithm should be able to quickly follow changes in the partner's behaviour.

All these results demonstrate the interplay of learning, clustering and logic within a probabilistic framework. This combination is important for trust development as it is for other cognitive abilities (e.g. Kemp et al., 2006). This is the strength of the presented algorithms. They are also suitable to better understand the Enfident Model, because the generative model of the algorithms matches with the visualisation of the Enfident Model. On the downside, the algorithms need much data and much computation time to converge to the target distributions. This gets even worse for relational models with more attributes and hidden variables. In addition, the algorithms disregard the temporal distance between two interactions. The transition probabilities between two connected state variables are all the same. But following the Enfident Model, a higher temporal distance should lead to a higher uncertainty about the next state value. For these reasons and some more mentioned in Section 7.4, I am not completely satisfied with the presented algorithms. They are subject to future research.

10 Conclusion

This dissertation contributes to a theory of trust with a conceptual and application-independent trust model and a set of accompanying requirements. This theoretical work is based on a thorough interdisciplinary discussion of trust in social and technical sciences; and it is complemented with modern algorithms from the field of statistical relational learning. For the evaluation, the Enfident Model is applied to a vehicular network scenario. The simulation features a virtual society of vehicles, which cooperate by exchanging interesting information. The evaluation shows that the postulated requirements and the Enfident Model together with the proposed algorithms lead to intuitive, reasonable and consistent results.

The next section summaries this dissertation. Section 10.2 then relates the Enfident Model back to interpersonal trust. The final section opens the view for further interesting research issues in the context of trust for technical systems.

10.1 Summary

In this dissertation, I analysed the definitions of trust in several disciplines. The analysis revealed different notions of trust, which I named probabilistic trust, interest-related trust, trust-related decision and trust-related behaviour. All these notions of trust address the uncertainty about another's behaviour. They target at inter-machine trust and not at interpersonal trust. The definitions abstract from interpersonal trust and the emotions that come along with it. Each is reduced so far that it refers to just one technical concept. This sequence of defined trust concepts is new in the literature.

The requirements then consider the mechanism that is necessary to compute these trust concepts. They focus especially on probabilistic trust, because the other trust concepts can be derived from it. The requirements identify base functionalities that are necessary to realise a trust mechanism. For example, a trust algorithm should generalise over all past trust situations to predict the trust for a new, unknown situational setting. The requirements are independent from an application and an implementation. In this way, they can guide in applying trust on any specific scenario. Systematic requirements for inter-machine trust are new in the literature.

The Enfident Model combines the required functionalities in one model, which features different views on trust. One view defines the entities of a trust situation: the cooperation partner, the bargaining situation and the task to perform. Various application-specific attributes describe the entities. They influence the trust in a specific situation. This variety of attributes is necessary to understand the trust situation and to predict the cooperation outcome well. The Enfident Model proposes the most complete assessment of the trust situation in the literature. In addition, it unifies trust models based on ratings of cooperation outcomes and those based on beliefs about the partner's characteristics in a coherent view (see Sections 4.2, 6.2 and 10.2.2).

Another view of the Enfident Model highlights the temporal aspects of trust. Trust develops over a temporal sequence of acts of cooperation. Each act of cooperation consists of a temporal sequence of interactions. Both temporal sequences introduce additional relations between the entities mentioned above. The trustor can evaluate trust at any time during an act of cooperation.

The view on the reasoning process conceptualises the attributes of an entity as the observable expression of an internal state. Each entity has its own internal state, which reflects its inner way of working. A state can change over time. For example, a software virus can change a robot's inner way of working. The evaluation showed that this conception of an entity helps

- to learn the specific trust for a certain situational setting;

- to take advantage of similarities between situational settings to judge a new, unknown setting;

- and to react on changes in the cooperation partner's behaviour.

The integration of these important trust functionalities in one model is unique in the literature. Moreover the reasoning process is based on experiences from the past, but also on a logical evaluation of present situational constraints (something like plausibility).

In addition, the Enfident Model contains a querying mechanism. It emphasises that trust can be assessed in different forms. For example, the Enfident Model can provide the trust for a specific act of cooperation, the trust in a specific cooperation partner and the trust as a general expectation about trust situations (see Section 10.2.3).

To implement the Enfident Model, I proposed modern algorithms for relational dynamic Bayesian networks. They combine logic, clustering, learning and reasoning based on probability theory to realise the postulated features of a trust mechanism.

I evaluated the Enfident Model with a virtual society of cognitive vehicles that cooperate in a vehicular network. This scenario emphasises that trust addresses a social problem. It is directed towards another system. The evaluation illustrates some requirements and some aspects of the Enfident Model. It justifies them intuitively with reasonable and consistent results.

The chain of technical thoughts follows a loop from the use cases and requirements to the Enfident Model and the implementation up to the evaluation. The evaluation results are then related back to the requirements. In contrast, the loop for the interdisciplinary considerations has not been closed completely yet. Therefore the next section sums up interesting connections between interpersonal trust and the Enfident Model.

10.2 The Enfident Model and Interpersonal Trust

This section shows selected parallels and differences between inter-personal and inter-machine trust. It closes the interdisciplinary discussion so far and justifies some aspects of the Enfident Model on a theoretical level.

10.2.1 Influencing Factors

Following the literature in the social sciences, the trusting person considers the trustee and the situational context to judge the trust situation (Section 3.1). The Enfident Model further distinguishes two context types: the situational context of the bargaining and that of the task execution. Moreover it conceptualises sequences of interactions (Section 6.1). There may be several situations of bargaining and task execution. This way of modelling details the concept from the social sciences for a technical application. It does not contradict that concept.

In addition, Gennerich emphasises that only the combination of all attributes together, attributes of the trustee and of the situational context, determines trust as described in Section 3.1. A person may be trustworthy regarding one task but not regarding another task. And additional information during the cooperation negotiation may place the cooperation in a different light. The Enfident Model agrees with this view. Its three entities influence the prediction only together, not independently.

10.2.2 Perceived Characteristics of the Cooperation Partner

Interpersonal trust depends on some attitudes towards the cooperation partner, like competence, consistency, promise fulfilment, and loyalty (see also Section 3.2.2). How does this observation relate to the Enfident Model as it does not contain such characteristics explicitly?

These characteristics are attributed to the cooperation partner based on past interactions and on recommendations of others. They can be understood as summarising descriptions of certain kinds of cooperation outcomes. If the cooperation partner, for example, devotes himself to someone in certain situations, he is said to be loyal to that one. Similarly if a professional does a good job in certain situations, he is called competent regarding these situations.

Consistency is different though. Someone can be consistently honest, consistently competent, consistently loyal, and so on. Consistency is more related to the form of the outcome distribution, while other characteristics seem more like an expected value of the outcome. An inconsistent behaviour could be said to lead to a broad outcome distribution. Reputation is different too. Someone can have a distinct reputation for loyalty, competence and so on. Reputation represents the opinion of a community. This community can have an opinion about all distinct characteristics of a person, but also about the person as a whole.

These considerations show that the Enfident Model indirectly includes such characteristics of the cooperation partner. It evaluates the same experiences that lead to perceived characteristics like competence, promise fulfilment and consistency. This way, trust computed by the Enfident Model will strongly correlate with those attitudes, although they are no direct input parameter of the Enfident Model. In fact, the researchers on interpersonal trust only investigated the correlations between trust and the characteristics. The results do not show the causality between them. For technical systems, I prefer that the algorithms have the full data instead of the derived attitudes.

Nonetheless thinking about these characteristics helps to identify all the properties of the cooperation outcome that are helpful for the trust development. *So the trustee's characteristics may be considered a development tool to understand the evaluation of the trust situation.* For technical systems, those groups of characteristics that regard professional relations (see Section 3.4.2) might be most relevant. They include, for example, expertise, fairness, honesty, integrity, promise fulfilment, discreteness and reputation. The Enfident Model emphasises this view by explicitly mentioning the ability and the willingness of the cooperation partner. This is in line with the work of Castelfranchi and Falcone (Section 6.2).

10.2.3 Trust as a Generalised and Specific Expectancy

Section 3.4.1 introduces trust as a generalised and a specific expectancy. In contrast, the arguments in this dissertation mainly aim at trust for a specific trust situation. Nonetheless the Enfident Model can produce a generalised expectancy as well – through the querying mechanism of Section 6.4. Trust can be computed for completely unknown entities. Then the result depends on the parameters of the Bayesian prior distribution – which is somehow comparable to Erikson's basic trust in form of an early disposition – and on an inductive inference over all previous situations – which is comparable to Rotter's idea of a learned generalised expectancy.

In between the completely generalised and completely specific expectancy, various trust scales have been developed for trust in intimate relationships, friends, professional guilds and so on, as described in Section 3.4.1 as well. Such expectancies could also be requested from the Enfident Model by inserting appropriate entities in the network. Thus one model can integrate all these notions of trust.

10.2.4 Trust as an Inner State or Manifest Behaviour

Chapter 2 points out that some scientists conceptualise interpersonal trust as cooperative manifest behaviour and others as an inner state. Modern researchers (e.g. Kassebaum, 2004) often unify both views in the way that they understand trust as an attitude with an affective, behavioural and cognitive expression.

This dissertation takes a step further by distinguishing probabilistic trust, interest-related trust, trust-related decision making and trust-related behaviour. Because trust-related decision making and behaviour depend also on aspects that are unrelated to trust, the Enfident Model ends right before the decision making process. It provides some input to the decision module, but does not say what should be done next or what information is correct or incorrect (for more details, see Section 4.3.2). It focuses on probabilistic and interest-related trust – the inner states of trust.

10.2.5 Summary

Comparing interpersonal trust with trust between cooperating technical systems helps to understand both of them. This is a central theme throughout this dissertation and a contribution to the state of the art. While some authors like Castelfranchi and Falcone (2010), Engler (2007), and Marsh (1994) regard some aspects of interpersonal trust, their interdisciplinary discussion omits many findings that are important for the presented work. The previous subsections show that interpersonal trust and the Enfident Model show some interesting parallels. They include aspects of what influences the trust development and aspects of what trust expresses. The parallels justify a few design aspects of the Enfident Model to some extent.

Besides these common aspects, trust between persons and trust between machines still have major differences. The reader should especially keep in mind that interpersonal trust is a phenomenon that is there, hidden in the person and, thus, subject to analysing investigation. In contrast, trust between technical systems is a mechanism synthesised to fulfil a certain purpose. And while a machine should act according to some rational principles, the research on interpersonal trust shows some interesting irrational findings (Section 3.4.3). Interpersonal trust also helps to constitute and maintain a person's own identity (Section 3.4.2). This is a far wider interpretation of the purpose than is necessary for cooperating machines.

10.3 Limitations of the Enfident Model and Future Research Issues

This dissertation marks just a small step towards trusting systems. Many important subjects are not covered. Society-level features, like those mentioned at the beginning of Chapter 4, are important for a social mechanism like trust. In some application scenarios, they even are a key for trust development.

Moreover the presented Enfident Model has some limitations, which should be investigated in the future, too. At present, the Enfident Model is more like a rough conceptual work piece, which still needs the finishing that comes from the variety of applications. Applying the Enfident Model on various real scenarios will bring up new insights and lead to improved algorithms. The proposed sampling schemas depict some features of the Enfident Model well, but they have some drawbacks as discussed in Section 9.5. There are other interesting techniques, which should be evaluated regarding trust development. The proposed requirements and the Enfident Model give some guidelines as to which features should be offered by a potential reasoning technique. This topic directly leads to benchmarks for trust algorithms. They mainly consist of data sets, which reflect an application scenario in a good quality. Evaluating reasoning techniques is only sensible with such benchmarks.

The topics above were mentioned several times in the dissertation. Some interesting limitations have not been touched though. At present, the proposed trust model offers no conclusive way to integrate experiences that come from observing the cooperation of others. All experiences are assumed to come from acts of cooperation between oneself and the cooperation partner. Because one self regards always the same entity, it is omitted in the Enfident Model. Consequently both cooperation partners must be modelled to include observed acts of cooperation. This raises many interesting questions like whether and how to model who is the trusting and the trusted system, and how to realise transfer learning for the various combinations of cooperation partners.

Furthermore the interaction between the agents as introduced in Section 6.1 could be modelled more extensively. In the Enfident Model, the state of an entity depends just on its previous state. For the cooperation partner this limitation could be too simplistic in some scenarios. Rather the partner's state could also depend on what the partner perceives. Consequently the causality could go from some observable attributes (the input attributes) to the current state variable and also from the preceding state variable to the current state variable. And from there, the causality could go to some other observable attributes (the output attributes). This results in a stochastic dynamic system (similar to a Kalman filter). While the Enfident Model omits the causality path from the input attributes to the current state variable, it still can react on changes of attributes through the inverse causality direction. For this reason, this extension of the Enfident Model could be unnecessary in many application scenarios.

The trust mechanism should support the system to act under high uncertainty. However it is only one part of this functionality. It must be complemented with trust-supported decision making (or information integration or control). Both, the trust computation and the decision making, could be combined in a single overall algorithm. This algorithm assesses the trust situation as the Enfident Model does and puts out a trust-related decision for an action. The inner trust state (probabilistic or interest-related trust) becomes irrelevant. For example, the information integration in a system could be implemented with a hidden Markov model and the Enfident Model with a dynamic Bayesian network (see Bishop, 2007 for an introduction on these techniques). Both models could be entangled in one big dynamic Bayesian network, which contains no inner trust state. The new algorithm would directly compute the integrated information without the detour of a trust value or trust distribution.

The trustee could be investigated and conceptualised, too. This is a rarely considered subject in the literature of the social sciences as well as the technical fields. It could give insights on how a trustee should decide, why he may betray and what could let him comply. This research would investigate good behaviour in the same way as malicious systems and the forces behind them. The model of the trustee could be combined with that of the trustor. This would result in an interesting overall interaction model for trust research, in which the alternating behaviour of both, the trustor and the trustee, could be investigated in various scenarios.

The research issues so far address trust between cooperating technical systems, as this is the subject of this dissertation. But certainly, the view could be opened to human-machine interaction: the technical system as a trustor, which needs the help of a human partner, and as a trustee, which is asked for cooperation by a person. While some aspects of trust in a human-machine setting might be similar to what is proposed in this dissertation, there might also be major differences as Chapter 3 and Section 10.2 emphasise. People do not act rationally. This leads to many interesting questions like: How does a person perceive a cognitive machine, which can act quite independently? What does a person expect from a machine in the role of a trustee? How will a person understand its role as a trusted cooperation partner, when he is asked for help by a machine? Trust in human-machine interaction is obviously a bit different and more complex than the rational concept of trust between technical systems. This was the main reason for me to start with technical systems as a first step.

Altogether I think, trust is an interesting research topic. The mixture of social and technical aspects is attractive and frequently brought up exciting findings and papers. The outlook in this section shows that the research on trust is still at the beginning. It promises many enhancements to trust theory and practice in the future. With a growing society of technical systems, more and more topics from the social sciences will have a technical relevance. Maybe somewhen, robots will even have a simple identity and self-conception. Trust could then support the preservation of the robot's identity.

Bibliography

E.R. Alexander, M.M. Helms, and R.D. Wilkins. The relationship between supervisory communication and subordinate performance and satisfaction among professionals. In *Public Personnel Management*, 18(4), pp. 415–429, 1989. ISSN 00910260.

G.E. Andrews, R. Askey, and R. Roy. *Special Functions*, volume 71 of *Encyclopedia of mathematics and its applications*. Cambridge University Press, Cambridge, UK, 1999. ISBN 0-521-78988-5.

C. Andrieu, N. de Freitas, A. Doucet, and M.I. Jordan. An introduction to MCMC for machine learning. In *Machine Learning*, 50, pp. 5–43, 2003. doi:10.1023/A:1020281327116.

Aristotle. *On Rhetoric: A Theory of Civic Discourse*. Oxford University Press, New York, 1991. ISBN 0-19-506486-0. Translated by G.A. Kennedy.

W. Bamberger, J. Schlittenlacher, and K. Diepold. A trust model for inter-vehicular communication based on belief theory. In *Social Computing (SocialCom), 2010 IEEE Second International Conference on*, pp. 73–80. 2010. doi:10.1109/SocialCom.2010.20. Best SocialCom Conference Paper Award.

W. Bamberger, J. Schlittenlacher, and K. Diepold. Ability-aware trust for vehicular networks. In *International Journal of Social Computing and Cyber-Physical Systems*, 1(3), pp. 286–307, 2012. doi:10.1504/IJSCCPS.2012.047701.

R. Bartoszyński and M. Niewiadomska-Bugaj. *Probability and Statistical Inference*. 2nd edition. Wiley, Hoboken, NJ, USA, 2008. ISBN 978-0-471-69693-3.

M. Behrisch, L. Bieker, J. Erdmann, and D. Krajzewicz. SUMO – simulation of urban mobility: An overview. In *SIMUL 2011, The Third International Conference on Advances in System Simulation*, pp. 55–60. IARIA, Wilmington, DE, USA, 2011. ISBN 978-1-61208-169-4.

D.J. Bem. Self-perception theory. In *Advances in Experimental Social Psychology*, 6, pp. 1–62, 1972. doi:10.1016/S0065-2601(08)60024-6.

H.W. Bierhoff. Vertrauen in Führungs- und Kooperationsbeziehungen. In A. Kieser (ed.), *Handwörterbuch der Führung*, number 10 in Enzyklopädie der Betriebswirtschaftslehre, pp. 2148–2158. Schäffer-Poeschel, Stuttgart, 1995. ISBN 3-7910-8043-1.

C.M. Bishop. *Pattern recognition and machine learning*. Springer, New York, NY, USA, 2007. ISBN 978-0387-31073-2.

K. Blomqvist. The many faces of trust. In *Scandinavian Journal of Management*, 13(3), pp. 271–286, 1997. doi:10.1016/S0956-5221(97)84644-1.

R.S. Burt and M. Knez. Kinds of third-party effects on trust. In *Rationality and Society*, 7(3), pp. 255–292, 1995. doi:10.1177/1043463195007003003.

V. Buskens. The social structure of trust. In *Social Networks*, 20(3), pp. 265–289, 1998. doi:10.1016/S0378-8733(98)00005-7.

M. Buss, M. Beetz, and D. Wollherr. CoTeSys — cognition for technical systems. In *International Journal of Assistive Robotics and Mechatronics*, 8(4), pp. 25–36, 2007. URL http://www.lsr.ei.tum.de/research/publications.

J.K. Butler. Reciprocity of dyadic trust in close male-female relationships. In *The Journal of Social Psychology*, 126(5), pp. 579–591, 1986. doi:10.1080/00224545.1986.9713630.

J.K. Butler. Towards understanding and measuring conditions of trust: evolution of a conditions of trust inventory. In *Journal of Management*, 17(3), pp. 643–663, 1991. doi:10.1177/014920639101700307.

T. Calvo, A. Kolesárová, M. Komorníková, and R. Mesiar. Aggregation operators: properties, classes and construction methods. In *Aggregation Operators*, pp. 3–104. Physica-Verlag, Heidelberg, Germany, 2002. ISBN 3-7908-1468-7.

CAR 2 CAR Communication Consortium. Manifesto. Technical report, version 1.1, August 2007. URL http://www.car-to-car.org/index.php?id=31.

C. Castelfranchi and R. Falcone. *Trust Theory: A Socio-Cognitive and Computational Model*. Wiley, Chichester, UK, 2010. ISBN 978-0-470-02875-9.

P.P.S. Chen. The entity-relationship model—toward a unified view of data. In *ACM Transactions on Database Systems*, 1(1), pp. 9–36, 1976. doi: 10.1145/320434.320440.

W. Chu, V. Sindhwani, Z. Ghahramani, and S.S. Keerthi. Relational learning with Gaussian processes. In B. Schölkopf, J. Platt, and T. Hoffman (eds.), *Advances in Neural Information Processing Systems 19*, pp. 289–296. MIT Press, Cambridge, MA, 2007. URL http://books.nips.cc/nips19.html.

J.S. Coleman. *Foundations of Social Theory*. 3rd edition. Harvard University Press, Cambridge, MA, USA, 2000. ISBN 0-674-31226-0.

L.L. Couch and W.H. Jones. Measuring levels of trust. In *Journal of Research in Personality*, 31(3), pp. 319–336, 1997. doi:10.1006/jrpe.1997.2186.

eBay Inc. All about feedback. Online web site, 2013. URL http://pages.ebay.com/help/feedback/allaboutfeedback.html. Visited on 2013-01-24.

C.S. Eichler. *Solutions for Scalable Communication and System Security in Vehicular Network Architectures*. Dissertation, Technische Universität München, München, 2009. URL http://mediatum2.ub.tum.de/doc/736760/736760.pdf.

E. ElSalamouny, V. Sassone, and M. Nielsen. Hmm-based trust model. In P. Degano and J.D. Guttman (eds.), *Formal Aspects in Security and Trust:*

6th International Workshop, FAST 2009, Eindhoven, The Netherlands, November 5-6, 2009, Revised Selected Papers, volume 5983 of Lecture Notes in Computer Science, pp. 21–35. Springer, Berlin/Heidelberg, Germany, 2010. doi:10.1007/978-3-642-12459-4_3.

M. Engler. Fundamental models and algorithms for a distributed reputation system. Dissertation, Universität Stuttgart, 2007. URL http://elib.uni-stuttgart.de/opus/volltexte/2008/3401.

W. Enkelmann. FleetNet – applications for inter-vehicle communication. In Intelligent Vehicles Symposium, 2003. Proceedings. IEEE, pp. 162–167. 2003. doi:10.1109/IVS.2003.1212902.

E.H. Erikson. Identity: Youth and Crisis. 2nd edition. Norton, New York, NY, USA, 1968. ISBN 0-393-01069-4.

R. Falcone, G. Pezzulo, and C. Castelfranchi. A fuzzy approach to a belief-based trust computation. In R. Falcone, S. Barber, L. Korba, and M. Singh (eds.), Trust, Reputation, and Security: Theories and Practice, volume 2631 of Lecture Notes in Computer Science, pp. 73–86. Springer, Berlin, Germany, 2003. doi:10.1007/3-540-36609-1_7.

P. Fassheber, H.G. Niemeyer, and C. Kordowski. Methoden und Befunde der Interaktionsforschung mit dem SYMLOG-Konzept am Institut für Wirtschafts- und Sozialpsychologie Göttingen. Report 18, Institut für Wirtschafts- und Sozialpsychologie der Georg-August-Universität Göttingen, Göttingen, Oktober 1990.

R.H. Fazio and R.E. Petty (eds.). Attitudes: Their Structure, Function, and Consequences. Psychology Press, New York, 2008. ISBN 978-1-84169-009-4.

T.S. Ferguson. A Bayesian analysis of some nonparametric problems. In The Annals of Statistics, 1(2), pp. 209–230, March 1973. doi:10.1214/aos/1176342360.

J.P. Forgas. *Interpersonal Behaviour: the Psychology of Social Interaction.* Pergamon Press, Sydney, 1985. ISBN 0-08-029868-0.

B.A. Frigyik, A. Kapila, and M.R. Gupta. *Introduction to the Dirichlet Distribution and Related Processes.* Technical Report UWEETR-2010-0006, Department of Electrical Engineering, University of Washington, Seattle, WA, USA, December 2010. URL https://www.ee.washington.edu/techsite/papers/refer/UWEETR-2010-0006.html.

B. Fristedt and L. Gray. *A Modern Approach to Probability Theory.* Birkhäuser, Cambridge, MA, USA, 1997. ISBN 0-8176-3807-5.

C. Gennerich. *Vertrauen: ein beziehungsanalytisches Modell – untersucht am Beispiel der Beziehung von Gemeindegliedern zu ihrem Pfarrer.* Huber, Bern, 2000. ISBN 3-456-83496-9.

E. Gerck. Trust as qualified reliance on information. In The COOK Report on Internet Protocol. Online periodical by COOK Network Consultants, pp. 19-24, January 2002. URL http://cookreport.com/. Last visited 2013-01-27.

L. Getoor and B. Taskar (eds.). *Introduction to Statistical Relational Learning.* MIT Press, Cambridge, MA, USA, 2007. ISBN 978-0-262-07288-5.

P. Golle, D. Greene, and J. Staddon. Detecting and correcting malicious data in VANETs. In *VANET '04: Proceedings of the 1st ACM International Workshop on Vehicular Ad Hoc Networks*, pp. 29–37. ACM, New York, NY, USA, 2004. doi:10.1145/1023875.1023881.

M.B. Gurtman and C. Lion. Interpersonal trust and perceptual vigilance for trustworthiness descriptors. In *Journal of Research in Personality*, 16(1), pp. 108–117, 1982. doi:10.1016/0092-6566(82)90044-7.

D. Heckerman, C. Meek, and D. Koller. Probabilistic entity-relationship models, PRMs, and plate models. In (Getoor and Taskar, 2007), pp. 201–218.

U. Heisig. Vertrauensbeziehungen in der Arbeitsorganisation. In (Schweer, 1997), pp. 121–153.

G. Hertkorn. *Mikroskopische Modellierung von zeitabhängiger Verkehrs-nachfrage und von Verkehrsflußmustern.* Dissertation, Universität zu Köln, Cologne, Germany, 2004. URL http://elib.dlr.de/21014/.

S. Hirche. Multi-PCA. Results of the brainstorming session for CoTeSys II on June 2nd, 2010. This document is an unpublished session protocol. There is no published material about this subject yet.

ISO/IEC11889-1. *International Standard ISO/IEC 11889-1. Information technology — Trusted Platform Module — Part 1: Overview.* ISO/IEC, Geneva, Switzerland, 1st edition, May 2009. URL http://standards.iso.org/ittf/PubliclyAvailableStandards.

ITU. *X.509: Information technology – Open systems interconnection – The Directory: Public-key and attribute certificate frameworks.* International Telecommunication Union (ITU), Geneva, Switzerland, November 2008.

A.J. Jones. On the concept of trust. In *Decision Support Systems*, 33(3), pp. 225–232, 2002. doi:10.1016/S0167-9236(02)00013-1.

A. Jøsang and R. Ismail. The beta reputation system. In *15th Bled Electronic Commerce Conference.* 2002. URL http://persons.unik.no/josang/papers/JI2002-Bled.pdf.

A. Jøsang, S. Marsh, and S. Pope. Exploring different types of trust propagation. In *Trust Management: 4th International Conference, iTrust 2006, Pisa, Italy, May 16–19, 2006,* pp. 179–192. Springer, Berlin, 2006. doi:10.1007/11755593_14.

U.B. Kassebaum. *Interpersonelles Vertrauen: Entwicklung eines Inventars zur Erfassung spezifischer Aspekte des Konstrukts.* Dissertation, Universität Hamburg, Hamburg, 2004. URL http://www.sub.uni-hamburg.de/opus/volltexte/2004/2125.

H.W. Kee and R.E. Knox. Conceptual and methodological considerations in the study of trust and suspicion. In *Journal of Conflict Resolution*, 14(3), pp. 357–366, 1970. doi:10.1177/002200277001400307.

C. Kemp, J.B. Tenenbaum, T.L. Griffiths, T. Yamada, and N. Ueda. Learning systems of concepts with an infinite relational model. In *Proceedings of the 21st National Conference on Artificial Intelligence – Volume 1 (AAAI'06)*, pp. 381–388. The AAAI Press, 2006. ISBN 978-1-57735-297-6.

D.A. Kenny. *Interpersonal Perception: a Social Relations Analysis.* Guilford Press, New York, 1994. ISBN 0-89862-114-3.

M. Koller. Risk as a determinant of trust. In *Basic and Applied Social Psychology*, 9(4), pp. 265–276, 1988. doi:10.1207/s15324834basp0904_2.

M. Koller. Psychologie interpersonalen Vertrauens. In (Schweer, 1997), pp. 13–26.

M. Kranz. Foundations of and applications for car-to-car communication. In *RADCOM 2008 Radar, Communication and Measurement.* GEROTRON COMMUNICATION GmbH, Martinsried, Germany, 2008. URL http://elib.dlr.de/53824.

A. Kröller, D. Pfisterer, C. Buschmann, S.P. Fekete, and S. Fischer. Shawn: A new approach to simulating wireless sensor networks. In *Proceedings of the 2005 Design, Analysis, and Simulation of Distributed Systems Symposium (DASD'05)*, pp. 117–124. 2005. ISBN 978-1-56555-294-4.

R.D. Laing, H. Phillipson, and A.R. Lee. *Interpersonal Perception: a Theory and a Method of Research.* Tavistock Publications, London, 1966.

J.L. Loomis. Communication, the development of trust, and cooperative behavior. In *Human Relations*, 12(4), pp. 305–315, 1959. doi:10.1177/001872675901200402.

N. Luhmann. *Trust and Power.* Wiley, Chichester, 1979. ISBN 0471997587. The original version is Luhmann, 1989.

N. Luhmann. *Soziale Systeme: Grundriß einer allgemeinen Theorie.* 2nd edition. Suhrkamp, Frankfurt am Main, 1985. ISBN 3-518-57700-4. For an English translation see Luhmann, 1995.

N. Luhmann. *Vertrauen: ein Mechanismus der Reduktion sozialer Komplexität*. 3rd edition. Enke, Stuttgart, 1989. ISBN 3-432-83773-9. For an English translation see Luhmann, 1979.

N. Luhmann. *Social Systems*. Stanford University Press, Stanford, 1995. ISBN 0-8047-1993-4. The original version is Luhmann, 1985.

D.J.C. MacKay and L.C.B. Peto. A hierarchical Dirichlet language model. In *Natural Language Engineering*, 1(03), pp. 289–308, 1995. doi:10.1017/S1351324900000218.

C.E. Manfredotti. *Modeling and Inference with Relational Dynamic Bayesian Networks*. Dissertation, Università degli Studi di Milano-Bicocca, Italy, 2009. URL http://boa.unimib.it/handle/10281/7829.

S.P. Marsh. *Formalising Trust as a Computational Concept*. Ph.D. thesis, University of Stirling, Scotland, 1994. URL http://users.xplornet.com/~spm/SteveMarsh/Publications.html.

K. Matheus, R. Morich, and A. Lübke. Economic background of car-to-car communication. In *Proceedings of the 2nd Braunschweiger Symposium Informationssysteme für mobile Anwendungen (IMA 2004)*. Braunschweig, Germany, 2004.

D.H. McKnight and N.L. Chervany. What trust means in e-commerce customer relationships: An interdisciplinary conceptual typology. In *International Journal of Electronic Commerce*, 6(2), pp. 35–59, 2001. ISSN 10864415.

G.H. Mead. *Mind, Self, and Society: From the Standpoint of a Social Behaviorist*. 19th edition. The University of Chicago Press, Chicago, IL, USA, 1974. ISBN 0-226-51668-7.

A.J. Menezes, P.C. van Oorschot, and S.A. Vanstone. *Handbook of Applied Cryptography*. 5th edition. CRC Press, August 2001. ISBN 0-8493-8523-7.

E.F. Nakamura, A.A.F. Loureiro, and A.C. Frery. Information fusion for wireless sensor networks: Methods, models, and classifications. In *ACM Computing Surveys*, 39(3), p. 9/1–9/55, 2007. doi:10.1145/1267070.1267073.

C. Narowski. *Vertrauen: Begriffsanalyse und Operationalisierungsversuch; Prolegomena zu einer empirischen psychologisch-pädagogischen Untersuchung der zwischenmenschlichen Einstellung: Vertrauen*. Dissertation, Eberhard-Karls-Universität, Tübingen, 1974.

R.M. Neal. *Bayesian Mixture Modeling by Monte Carlo Simulation*. Technical Report CRG-TR-91-2, University of Toronto, June 1991. URL http://www.cs.toronto.edu/~radford/bmm.abstract.html.

R.M. Neal. Markov chain sampling methods for Dirichlet process mixture models. In *Journal of Computational and Graphical Statistics*, 9(2), pp. 249–265, 2000. ISSN 1061-8600.

OpenStreetMap Project. OpenStreetMap. Online web site, 2013. URL http://www.openstreetmap.org. Last visited on 2013-01-24.

M.E. Oswald. Bedingungen des Vertrauens in sozialen Situationen. In M.K.W. Schweer (ed.), *Vertrauen und soziales Handeln*, pp. 78–98. Luchterhand, Neuwied, Germany, 1997. ISBN 3-472-02990-0.

S.J. Pan and Q. Yang. A survey on transfer learning. In *IEEE Transactions on Knowledge and Data Engineering*, 22(10), pp. 1345–1359, 2010. doi:10.1109/TKDE.2009.191.

C.D. Parks, R.F. Henager, and S.D. Scamahorn. Trust and reactions to messages of intent in social dilemmas. In *Journal of Conflict Resolution*, 40(1), pp. 134–151, 1996. doi:10.1177/0022002796040001007.

J. Pearl. *Probabilistic Reasoning in Intelligent Systems: Networks of Plausible Inference*. Morgan Kaufmann, San Mateo, CA, USA, 1988. ISBN 0-934613-73-7.

J. Pearl. *Causality: Models, Reasoning, and Inference*. Cambridge University Press, Cambridge, UK, 2000. ISBN 978-0-521-77362-1.

F. Petermann. *Psychologie des Vertrauens*. 3rd edition. Hogrefe, Göttingen, 1996. ISBN 3-8017-0932-9.

J. Pitman. *Combinatorial Stochastic Processes*, volume 1875 of *Lecture Notes in Mathematics*. Springer, Berlin/Heidelberg, Germany, 2006. doi: 10.1007/b11601500.

S.D. Ramchurn, D. Huynh, and N.R. Jennings. Trust in multi-agent systems. In *The Knowledge Engineering Review*, 19(1), pp. 1–25, 2004. doi:10.1017/S0269888904000116.

S.D. Ramchurn. *Multi-Agent Negotiation using Trust and Persuasion*. Ph.D. thesis, University of Southampton, UK, 2004. URL http://eprints.ecs.soton.ac.uk/10200/.

M. Raya. *Data-Centric Trust in Ephemeral Networks*. Ph.D. thesis, École Polytechnique Fédérale de Lausanne, Switzerland, June 2009.

M. Raya, P. Papadimitratos, V. Gligor, and J.P. Hubaux. On data-centric trust establishment in ephemeral ad hoc networks. In *INFOCOM 2008. The 27th Conference on Computer Communications. IEEE*, pp. 1238–1246. April 2008. doi:10.1109/INFOCOM.2008.180.

J.K. Rempel, J.G. Holmes, and M.P. Zanna. Trust in close relationships. In *Journal of Personality and Social Psychology*, 49(1), pp. 95–112, 1985. doi:10.1037/0022-3514.49.1.95.

A. Rettinger, M. Nickles, and V. Tresp. A statistical relational model for trust learning. In Padgham, Parkes, Müller, and Parsons (eds.), *Proc. of 7th Int. Conf. on Autonomous Agents and Multiagent Systems (AAMAS, 2008)*, pp. 763–770. International Foundation for Autonomous Agents and Multiagent Systems, Richland, SC, May 2008. ISBN 978-0-9817381-1-6.

G. Ropohl. *Eine Systemtheorie der Technik: Zur Grundlegung der Allgemeinen Technologie*. Hanser, München, 1979. ISBN 3-446-12801-8.

J.B. Rotter. Interpersonal trust, trustworthiness, and gullibility. In *American Psychologist*, 35(1), pp. 1–7, 1980. doi:10.1037/0003-066X.35.1.1.

J. Sabater and C. Sierra. Reputation and social network analysis in multi-agent systems. In *AAMAS '02: Proceedings of the First International Joint Conference on Autonomous Agents and Multiagent Systems*, pp. 475–482. ACM, New York, NY, USA, 2002. doi:10.1145/544741.544854.

K.D. Schmidt. *Maß und Wahrscheinlichkeit*. Springer, Berlin, 2009. ISBN 978-3-540-89729-3.

M. Schweer (ed.). *Interpersonales Vertrauen: Theorien und empirische Befunde*. Westdeutscher Verlag, Opladen, Germany, 1997. ISBN 3-531-13033-1.

S. Sen and P.S. Dutta. The evolution and stability of cooperative traits. In *Proceedings of the First International Joint Conference on Autonomous Agents and Multiagent Systems: Part 3*, pp. 1114–1120. ACM, New York, NY, USA, 2002. doi:10.1145/545056.545081.

G. Shafer. *A mathematical theory of evidence*. Princeton University Press, Princeton, NJ, 1976. ISBN 0-691-08175-1.

G. Simmel. *Soziologie: Untersuchungen über die Formen der Vergesellschaftung*. 5th edition. Dunker & Humbolt, Berlin, 1968.

S.B. Sitkin and N.L. Roth. Explaining the limited effectiveness of legalistic "remedies" for trust/distrust. In *Organization Science*, 4(3), pp. 367–392, 1993. URL http://www.jstor.org/stable/2634950.

C. Stiller, G. Färber, and S. Kammel. Cooperative cognitive automobiles. In *Intelligent Vehicles Symposium, 2007 IEEE*, pp. 215–220. June 2007. doi:10.1109/IVS.2007.4290117.

M. Strack. *Sozialperspektivität: Theoretische Bezüge, Forschungsmethodik und wirtschaftliche Praktikabilität eines beziehungsdiagnostischen Konstrukts*. Universitätsverlag Göttingen, Göttingen, Germany, 2004. ISBN 3-930457-63-6.

H. Strasser and S. Voswinkel. Vertrauen im gesellschaftlichen Wandel. In (Schweer, 1997), pp. 217–236.

L.H. Strickland. Surveillance and trust. In *Journal of Personality*, 26(2), pp. 200–215, 1958. doi:10.1111/j.1467-6494.1958.tb01580.x.

B. Taskar, P. Abbeel, M.F. Wong, and D. Koller. Relational Markov networks. In (Getoor and Taskar, 2007), pp. 175–199.

J.T. Tedeschi, S. Lindskold, J. Horai, and J.P. Gahagan. Social power and the credibility of promises. In *Journal of Personality and Social Psychology*, 13(3), pp. 253–261, 1969. doi:10.1037/h0028281.

Y.W. Teh, M.I. Jordan, M.J. Beal, and D.M. Blei. Hierarchical Dirichlet processes. In *Journal of the American Statistical Association*, 101(476), pp. 1566–1581, 2006. doi:10.1198/016214506000000302.

R. Trappl (ed.). *Cybernetics: Theory and Application.* Springer, Berlin, 1983. ISBN 3-540-12548-5.

J. Van Gael. *Bayesian Nonparametric Hidden Markov Models.* Ph.D. thesis, University of Cambridge, Cambridge, UK, 2011.

J. Van Gael, Y. Saatci, Y.W. Teh, and Z. Ghahramani. Beam sampling for the infinite hidden Markov model. In *Proceedings of the 25th International Conference on Machine Learning*, pp. 1088–1095. ACM, New York, NY, USA, 2008. doi:10.1145/1390156.1390293.

M. Wooldridge. *Reasoning about Rational Agents.* MIT Press, Cambridge, MA, USA, 2000. ISBN 0-262-23213-8.

Z. Xu. *Statistical Relational Learning with Nonparametric Bayesian Models.* Dissertation, LMU München: Fakultät für Mathematik, Informatik und Statistik, Munich, Germany, 2007. URL http://nbn-resolving.de/urn:nbn:de:bvb:19-76196.

Z. Xu, K. Kersting, and V. Tresp. Multi-relational learning with Gaussian processes. In *Proceedings of the 21st International Joint Conference on Artifical intelligence*, pp. 1309–1314. Morgan Kaufmann Publishers Inc., San Francisco, CA, USA, 2009. URL http://www.aaai.org/ocs/index.php/IJCAI/IJCAI-09/paper/view/648.

B. Yu and M.P. Singh. Distributed reputation management for electronic commerce. In *Computational Intelligence*, 18(4), pp. 535–549, 2002. doi:10.1111/1467-8640.00202.

G. Zacharia and P. Maes. Trust management through reputation mechanisms. In *Applied Artificial Intelligence*, 14(9), pp. 881–907, 2000. doi:10.1080/08839510050144868.

A.M. Zak, J.A. Gold, R.M. Ryckman, and E. Lenney. Assessments of trust in intimate relationships and the self-perception process. In *The Journal of Social Psychology*, 138(2), pp. 217–228, 1998. doi:10.1080/00224549809600373.

W. Zhang, S.K. Das, and Y. Liu. A trust based framework for secure data aggregation in wireless sensor networks. In S.K. Das (ed.), *Proceedings of the 2006 3rd Annual IEEE Communications Society on Sensor and Ad Hoc Communications and Networks (SECON '06)*, volume 1, pp. 60–69. IEEE, September 2006. doi:10.1109/SAHCN.2006.288409.